D1387893

Principles of Semiconductor Network Testing

Principles of Semiconductor Network Testing

Amir Afshar

Butterworth–Heinemann

Boston Oxford Melbourne Singapore Toronto Munich New Delhi Tokyo

Recognizing the importance of preserving what has been written, Butterworth–Heinemann prints its books on acid-free paper whenever possible.

Library of Congress Cataloging-in-Publication Data

Afshar, Amir.

Principles of semiconductor network testing / Amir Afshar.

p. cm.

Includes bibliographical references and index.

ISBN 0-7506-9472-6 (hardcover: alk. paper)

1. Integrated circuits—Testing. 2. Semiconductors—Testing.

I. Title.

TK7874.A339 1995

621.3815′48—dc20

95-13386

CIP

British Library Cataloguing-in-Publication Data

A catalogue record for this book is available from the British Library.

The publisher offers discounts on bulk orders of this book.
For information, please write:

Manager of Special Sales
Butterworth–Heinemann
313 Washington Street
Newton, MA 02158-1626

10 9 8 7 6 5 4 3 2 1

Printed in the United States of America

To my wonderful mother and sweet daughter Jessica Shireen

Contents

Foreword

Back in the good old days, the semiconductor test industry consisted of a technician at his bench with a trusty Simpson 260 VOM and a trimpot. The semiconductor designs were "tossed over the wall" from design to manufacturing, and the process and test groups did what they could to get products out the door. Each went his own way, often and inadvertently at cross purposes, but the demand for semiconductors was great, complexity was low—mainly compressions of discrete designs onto a silicon substrate—and the semiconductor house shipped what it could and the customer took what he could get.

With the advent of the minicomputer, the test technician's task became automated. The heart of the test system was still a VOM, a computer-controlled relay matrix simulated the technician's rapidly moving hands, and the CPU recorded readings, compared limits, and manipulated automated handling equipment. The human test technician evolved into a test engineer, who wrote the computer programs to control the testing operation. Manufacturing volume was greatly increased, and reliability was enhanced as subjective human judgments were removed from pass/fail criteria.

Looking back, the good old days of semiconductor test were really the dark ages. The test instrumentation and test methods then available limited many parameter measurements to little more than wiggle testing. Data sheets were full of "Guaranteed by Design," "Typical Values Only," "Periodic Lot Sampling" (right!) entries but very little meat on precise device performance values. The testing process was seen mainly as an impediment to the flow of semiconductors from the shipping dock. The designer, process engineer, and most importantly, the customer, gained little value from the testing process.

Today, the test industry is going nova. Witness the whole new integrated semiconductor design software packages that feature "DFT" (Design for Test), Boundary Scan, and fault coverage algorithms. State-of-the-art test equipment has computing power, accuracy, pin counts, and tester per pin architecture only dreamed of a few years ago. The test floor itself is an integral part of the design and process system, with the test equipment feeding quantities of statistical process control information and performance at full operating speed back to a central database. From this database, the quality and consistency of the fabrication process can be monitored on a real-time basis, and the design engineer gains valuable insight into real-world performance versus design goals and where he can next push the envelope. Armed with this information, the salesforce learn why their product is better than the competitor's.

Successful semiconductor companies are the ones that maintain this tightly coupled feedback loop between design, process, and test engineering. In today's competitive marketplace, no longer can a design be tossed over the wall. It is time to market, the performance edge, quality, and test floor productivity that determine who will be around tomorrow. That is why it is important for test issues to be settled early in the design process. That is also why it is of the utmost importance that the information that is collected and fed back from the testing process be of the highest quality. It does not one any good, and in fact a great deal of harm, if the test data is contaminated by uncertainty, faulty test methodology, inaccuracy, noise, and distortion.

This book addresses the fundamental issues underlying the semiconductor test discipline. Test engineers must understand the basic principles of semiconductor fabrication and processing and have an in-depth knowledge of circuit functions, instrumentation, and noise sources. In doing so, they will evolve toward greater professionalism and add value to the products being tested. This book has gathered under one cover diverse and comprehensive information, usually scattered far and wide, that the test and process professional will find invaluable. The techniques outlined herein will help insure that the test methods and data collected will reflect actual device performance, not "testing the tester" or being lost in the noise floor.

As a final note, the chapter on digital signal processing brings up an interesting development in the test world. Sometime in the early 1970s the test profession split into Digital and Analog camps and pretty much pursued their own separate ways. Now, with the pressure on for systems-on-a-chip solutions, analog and digital circuits are finding their way onto the same die, giving rise to the mixed-signal device. Just as a digital test engineer is learning that as data rates get faster all signals are analog, the analog test engineer is finding that it behoves him to get the signal measurements into the digital domain as quickly and as close to the part as possible. Once in the digital domain, the processing power of today's DP engines allows the engineer to realize noiseless filters, ideal true RMS voltmeters, lossless peak detectors, arbitrary waveform generators, and so on, that are limited only by mathematical precision and imagination, not tolerance, drift, and inaccuracy of physical components. The field of DSP-based testing is welding the split back together.

Hey, anybody see my Simpson 260 ...?

DONALD E. GREER
Senior Test Engineer

Preface

Intent

This book provides comprehensive information about semiconductor microcircuit test procedures and is recommended for all levels of semiconductor product and test engineers, from entry level and up. It can also be used by skilled test technicians since it includes step-by-step test procedures.

The ultimate challenge for a complicated microcircuit design is the functional verification of the design by testing. In this book you will find all the information needed to perform this verification.

Engineering schools teaching semiconductor-related subjects can use this book as a text or as complementary information.

The Book in Brief

This book contains eight chapters. The first two chapters provide a foundation for understanding diode and transistor operation and a brief description of the semiconductor process from silicon ore to processed wafer and implanted microcircuits. Methods of testing a die on a wafer, along with an explanation of the wafer test hardware, are presented with test and product engineers in mind.

The chapter on noise identification contains the most vital information for semiconductor test and product engineers. The methods for identifying noise and suggestions for suppressing it are practical, nonmathematical, and easy to comprehend. In another chapter digital signal processing, a new and progressive field of communication, is explained thoroughly and succinctly without complicated mathematical or theoretical content.

The remaining chapters provide in-depth descriptions of analog and digital devices and complete test procedures. These tests can be performed in a laboratory or they can be used as a guide for writing test programs.

Acknowledgments

Many friends have helped me by either providing technical references or contributing their expertise. I would like to acknowledge their help in providing the latest innovations in the field. Thanks to Don Sabin, Said Huq, Masood Kazemi, Diane Trevino, Lynn Plecque, Jennell Picard, and Salem Mansoori, all from National Semiconductor Corporation; Jayma Vaughan and Shona Kerly, from Intel Corporation; Kevin Kurtz from CerProbe Corporation; Vicki Benson from Lucid Information Center; and David Senum from Micro Component Technology. Special thanks go to good friends and fine engineers Don Greer and Janardhan Gottipati for their keen attention.

Amir Afshar

CHAPTER 1

Diode and Transistor Operation

Introduction

Any semiconductor integrated circuit (IC), regardless of size or technology, consists of hundreds to millions of diodes and transistors packed into a small space, usually less than 0.5 cm^2 in size—all operating together. This chapter presents practical information in simple, understandable terms. A variety of schematics and figures are provided to illustrate the information presented. Mathematical equations and formulas are minimal, unless their inclusion is necessary for clarity. The simplicity of the contents of this chapter should not be a basis of judgment of the contents of the whole book. Material covered in Chapter 2 and beyond, although based on the contents of this chapter, requires full attention, and in most cases the reader should review the material a number of times for full comprehension.

It is essential to master this chapter before going further. Within the text, the first use of the name of the device and/or component appears in full followed by its abbreviation or acronym.

Topics

1.1. Semiconductor materials; description and how they are used to create diodes and transistors. Methods of biasing and modes of operation of diodes are explained in a purely nonmathematical fashion without theories.

1.2. Bipolar transistors, including fundamental differences between P-N-P and an N-P-N transistors, emphasizing the parameters that are of interest to semiconductor product, test or design engineers, such as

- A characteristic curve of a TTL
- Current and voltage flow
- Common emitter (CE) configuration
- Common collector (CC) configuration

Figure 1.1 An example of comparison between a VLSI chip and a daisy petal. This particular chip contains over 150 000 diodes and transistors. (Courtesy of Intel Corp.)

- Operation modes (cut-off, saturation and linear regions)
- Bipolar inverter operation

1.3. MOS transistors, including

- Description, make-up, uses, FET, JFET, gates, Q points, source and drain, use of ohmmeter to find polarity
- CMOS inverter operation, including voltage and current transfer
- CMOS vs. TTL transistors
- Input and output characteristics of a TTL shown on an *I* vs. *V* coordinator, and how to identify the segments of each curve for test purposes

Since diodes and transistors are the major building blocks of electronic circuitry, an understanding of the concept of their operation is essential in developing comprehensive test software. The input and output (I/O) of a semiconductor device originates from a specified internal diode or transistor; therefore, test methods described in this book deal essentiall with the I/O of these elements.

Diodes and transistors are electronic elements made from processed semiconductor materials. A diode is the simplest device made from semiconductor materials. A transistor consists of two diodes back to back and is used as a linear amplifier in linear circuits or as an on/off switch in digital circuitry.

1.1 Semiconductor Materials

Semiconductors are made from either silicon or germanium. The characteristics of these materials fall somewhere between those of conductors and insulators. The resistance of these materials to electron conduction makes them ideal materials for transistors and diodes.

A pure semiconductor crystal has equal numbers of electrons and holes. Holes are sites that have lost their electrons and are seeking them; therefore, they act as positive charges in current conduction. Electrons act as negative particles.

Current flow in semiconductor material is determined by the number and location of holes and electrons; when they are equal, there is no current conductivity until an external voltage is applied. The number of hole–electron pairs created in semiconductors increases with temperature, causing a voltage drop. For example, as temperature increases the voltage across a diode decreases at a rate of about 2 mV/°C. Breakdown ensues at the threshold voltage for semiconductor diodes and transistors. This will be discussed in Section 1.1.2.

Semiconductors undergo several processes before they are ready for transistor implantation. These processes are explained briefly under the related topics in Chapter 2. Interested readers may also refer to any book on semiconductor processing for more complete information.

Two types of semiconductors, negative (N-type) and positive (P-type), are created by injecting nonsemiconductor substances called *impurities* into ordinary semiconductor material. These impurities are also called *dopes* or *dopants* and the process of adding impurities is called *doping.*

Phosphorus atoms used as dopant are known as *donors* because they offer electrons. P-type atoms are called *acceptors* because they accept electrons. Acceptor sites are also called *holes.* Boron atoms are usually used as the dopant for the creation of holes.

In P-type semiconductor creation, some electrons are removed from the semiconductor by the doping process to create more holes. In N-type creation, electrons are added to carry current. A diode is one of the simplest electronic devices. It is a rectifier with a defined boundary that separates the negative from the positive charges. A diode is formed when a P-type material (or anode) is joined with an N-type material (or cathode). The joining edge is called the *junction.* This junction creates a minor capacitance that affects the speed of operation. The junction capacitance will be discussed in Section 1.2.6.

The direction of current flow is determined by the excess electrons or holes on either side of the junction. The applied voltage causes the current flow.

A diode is prepared for use by *biasing.* Biasing can be either forward or backward (reverse). Figure 1.2 shows the diode symbol, along with the structure and a representation of the resistance to current flow in each side.

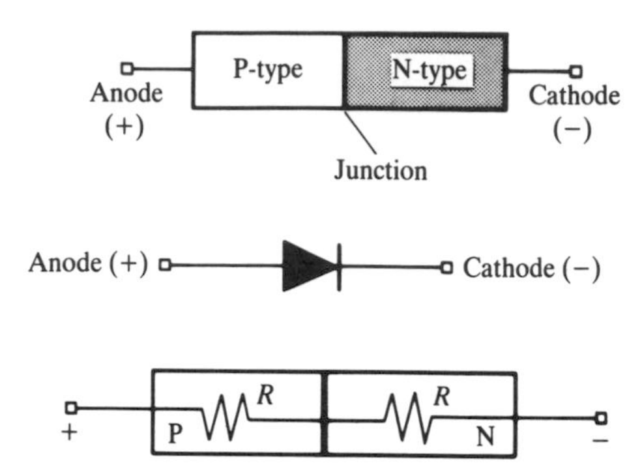

Figure 1.2 (From top) P-N junction, diode symbol, and representation of resistance to current flow in each side.

1.1.1 Diode Biasing

The voltage that is applied to a junction diode to create a flow of current is termed the biasing voltage. The direction of the current flow depends on the polarity of the biasing voltage. This applied voltage must be sufficient to override the resistance already existing in the junction diode that is depicted in Figure 1.2. An ideal diode is a perfect conductor when forward biased and a perfect insulator when reverse biased, like an on/off switch.

1.1.2 Forward Biasing

If the negative side of the power supply is connected to the cathode side of the diode and the positive side is connected to the anode side, electrons will move across the junction from the cathode side to the holes in the anode side. This current flow is forward biased and will continue as long as the power supply is sufficient. Forward biasing of a transistor affords the largest power gain with the least distortion; therefore, most linear signal amplifiers use bipolar transistors that are forward biased. Forward biasing and its related curve are represented in Figure 1.3.

The voltage drop across a semiconductor diode is dependent on the semiconductor material and independent of biasing voltage or other factors. The drop is about 0.7 V for silicon and about 0.3 V for germanium. This means that a germanium diode must have an applied voltage of at least 0.3 V, at room temperature, before it begins to conduct. For silicon, the voltage required is 0.7 V or more. This voltage, known as *junction electrostatic voltage drop*, is the amount of energy required to cause an electron to cross the junction between the N side and the P side.

The voltage at which the conduction of current begins is defined as the *threshold voltage*. Breakdown occurs at breakdown voltage, in which an unlimited amount of current can pass through the device (avalanche).

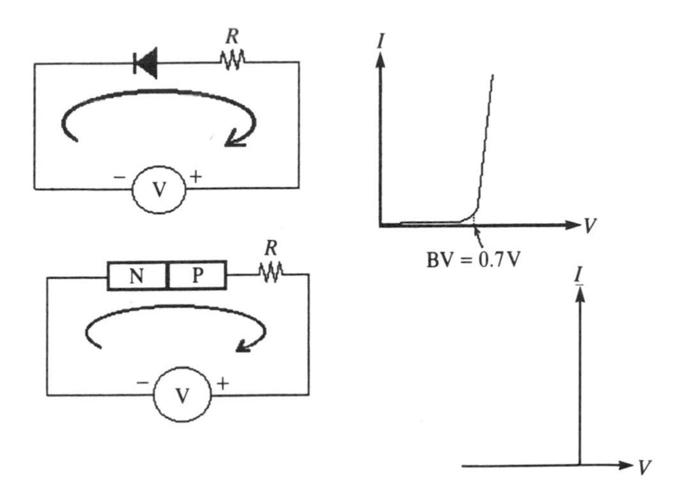

Figure 1.3 Forward biasing representative bias curve for an ideal diode.

There are a number of factors that must be considered when determining the magnitude of the current passing through a diode: diode material, circuit resistance, and the applied voltage magnitude.

Example A silicon diode, a 10 kΩ resistor, with a 10 V power supply in forward-bias mode will have a current given as follows.

$$10\text{ V (supply)} - 0.7\text{ V (drop across silicon)} = 9.3\text{ V}$$

$$V = RI$$

$$9.3 = 10\text{ k}\Omega \times I \quad \text{A}$$

$$I = 0.00186\text{ A or } 1.86\text{ mA}$$

For safety, the current-limiting resistor, 10 kΩ, should be employed in series. The size of the resistor limits the amount of current passing through the device. The larger the resistor, the less current gets through (Ohm's law).

1.1.3 Reverse Biasing

In reverse biasing, the negative side of the power supply is connected to the anode side of the diode and the positive side to the cathode. After electrons coming from the power supply have filled the holes in the P side, further electrons cross the junction and force electrons on the N side away from the junction. The electrons on each side of the junction repel each other creating a region that is free of either electrons or holes. This region is called the depletion region and there is no current flow. By shutting off current flow, the diode acts as an open switch. An insignificant amount of *leakage current* across the junction does occur. Consider Figure 1.4. The current will increase acutely (avalanche) at breakdown voltage (BV) with a small voltage increase.

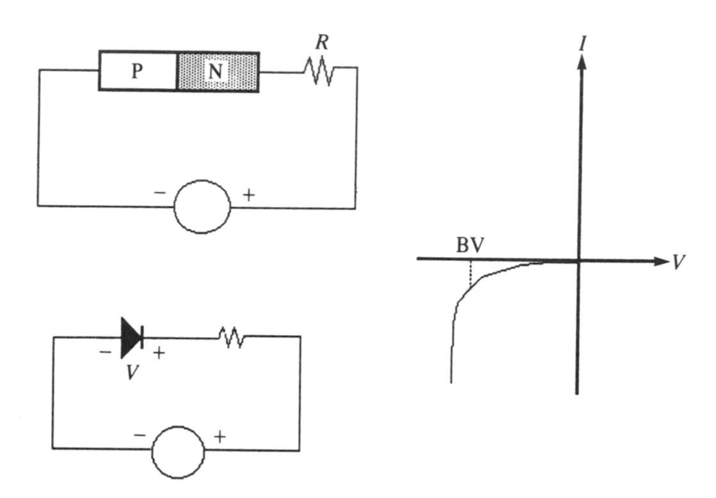

Figure 1.4 Reverse biased configuration.

1.1.4 Varieties of Diodes

There are many varieties of diodes and their use depends on their application. These devices comprise two terminals with characteristic behavior that is nonlinear with temperature.

A diode usually has an identification number that begins with 1N. A band or strip at the cathode end of a diode usually marks its polarity; this can be verified by an ohmmeter. If an ohmmeter reads very low when it is connected across a diode, the cathode of the diode is connected to the ground side of the ohmmeter (forward biased). The polarity of the diode is what is important here, and the resistance reading is not significant.

1.2 Bipolar Transistors

A bipolar transistor is composed of three semiconductors fused together to form two junctions. The three components of a transistor can be arranged to form two types of transistors: N-P-N or P-N-P. The middle area of the transistor, called the base, is narrower than the other two and contains either negative or positive charges. The other two parts of a transistor are called the emitter and the collector. The emitter is more heavily doped than the other regions and the emitter and collector cannot be interchanged without significantly changing the attributes of the device.

Figure 1.5 shows the two types of transistors and the arrangements of their sections.

The base and emitter of a P-N-P transistor form a P-N junction and the collector and base another P-N junction. The more heavily doped emitter creates an intensity of current within the device. These two junctions behave as if two junction diodes are connected back to back, as symbolized in Figure 1.5.

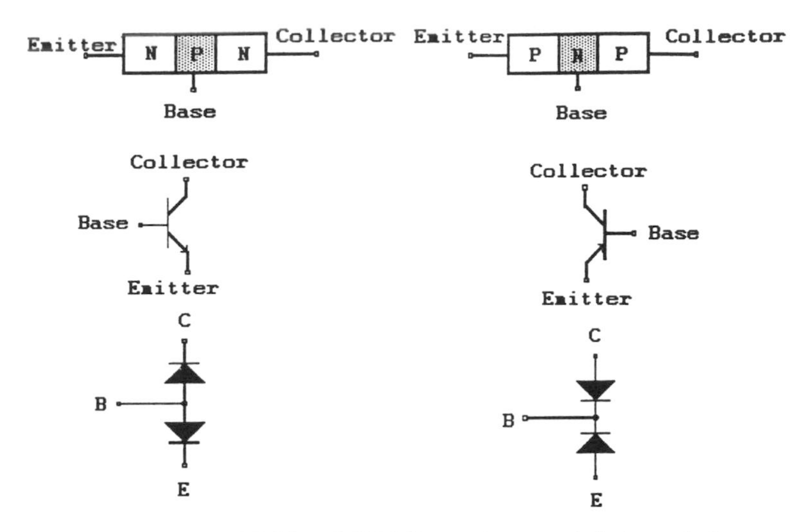

Figure 1.5 N-P-N and P-N-P transistors and their symbols.

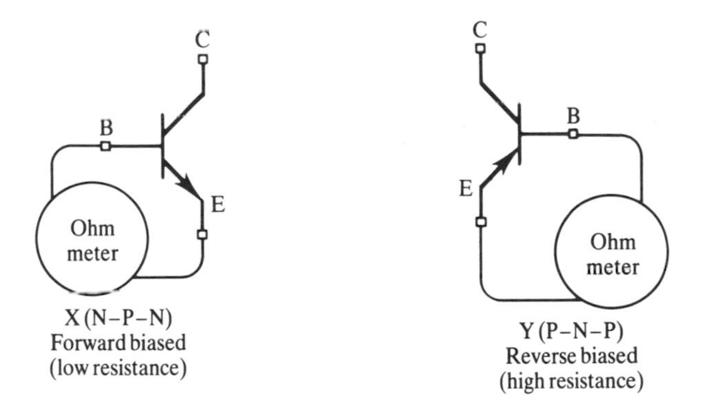

Figure 1.6 Biasing direction.

An ohmmeter can verify the direction of biasing in a transistor. If the terminals of an ohmmeter are connected to the base and emitter of transistors X and Y in Figure 1.6, low resistance will register in case X (forward biased) and high resistance in case Y (reverse biased).

An ohmmeter can also distinguish a bad transistor from a good one. In an operable transistor, the base/emitter (BE) and the base/collector (BC) junctions should demonstrate diode action, and the collector/emitter (CE) junction should read a high resistance. Figure 1.6 shows the connections. Both transistors have P-N junctions in common, with the same function as the junction diode. The arrow on the emitter in Figure 1.6 designates the type of transistor.

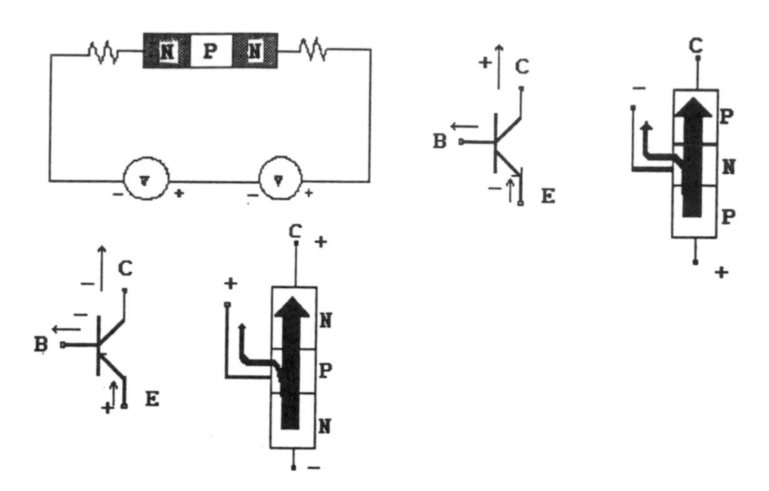

Figure 1.7 Biasing an N-P-N transistor.

The current direction in a transistor starts from the emitter, passing through the base and collector. The base current determines the amount of current flowing from the emitter to the collector. From this we can conclude that the base is the current-controlling section in a transistor. Holes and electrons are the current carriers in P-N-P and N-P-N transistors, respectively.

Transistors are called nonlinear because the relationship between the current and the voltage is not linear. They are current-controlled, meaning that a small increase in the input current causes a large change at the output. Forward biasing of the emitter/base (EB) junction and reverse biasing of the base/collector (BC) junction is the *only* arrangement that assures the operation of a transistor. The correct way to bias an N-P-N transistor is shown in Figure 1.7 (left); the direction of current in a P-N-P transistor and its representation are shown on the right.

1.2.1 Characteristic Curve of a TTL (Transistor–Transistor Logic)

The swing of the output voltage of a transistor is limited by the cutoff value at one extreme and the saturation value at the other. When the EB of a transistor is not forward biased, cutoff occurs. This means that there is no effect on the collector current from changes in the base/emitter (BE) voltage. Changes in the input current will have no effect on the output voltage when the collector/base (CB) is forward biased and saturation occurs.

The region between the cutoff and saturation is the *active region*, or the area within which a transistor must be operated. Only proper biasing can ensure that the transistor will operate in this region.

The best position for quiescent output voltage (Q point) is in the middle of the active region. In Figure 1.8 the quiescent output voltage is about 12 V. It can

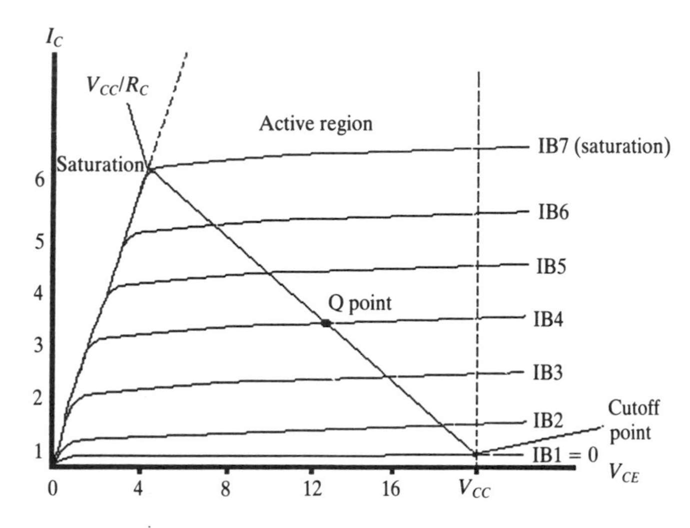

Figure 1.8 A common emitter TTL characteristic curve.

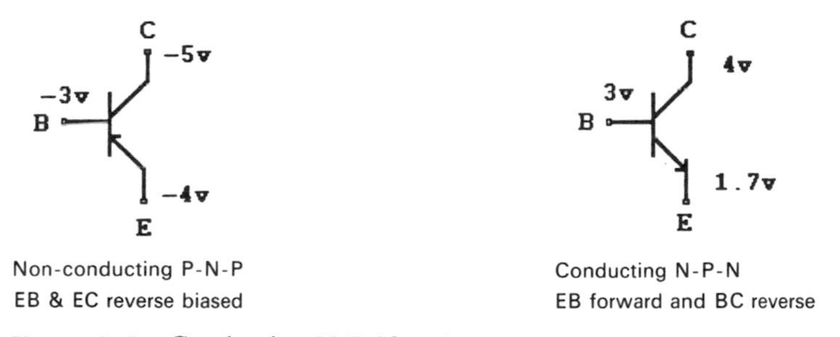

Figure 1.9 Conducting N-P-N and nonconducting P-N-P transistor.

swing from 5 to 20 V. The procedure for determining the location of the Q point, and the theoretical details of biasing, are outside of our concerns; the interested reader can find them in almost any electronics textbook.

By measuring the voltage at each element of the transistor, with polarities, it is easy to verify that the transistor is, in fact, conducting. The base of an N-P-N transistor must always be more positive than its emitter, because the EB is forward biased. Its collector must be more positive than the base, which means the (BC) junction is reverse biased. Figure 1.9 shows a conducting N-P-N and nonconducting P-N-P transistor with the required bias voltages.

1.2.2 Current Flow and Voltage

The majority of the current flow in a forward-biased N-P-N transistor consists of emitter current (I_E) and collector current (I_C). The base current (I_B) and the forward bias current form the emitter to the base are quite inconsequential. I_E is

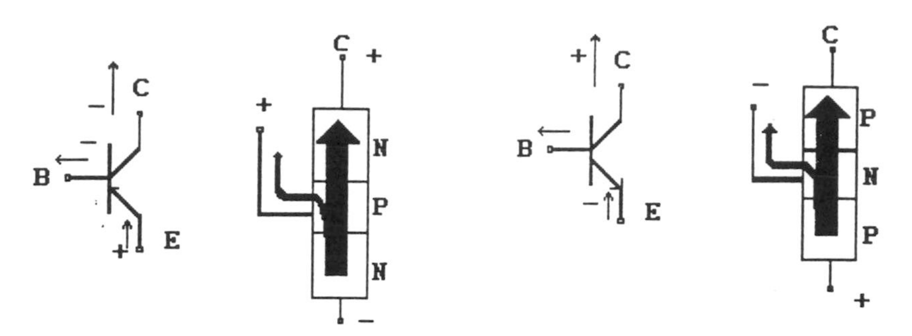

Figure 1.10 Current flow in a P-N-P- and an N-P-N- transistor.

actually I_B plus I_C:

$$I_E = I_C + I_B \tag{1}$$

Figure 1.10 shows the direction of these currents.

The base of a transistor is much narrower than the other two sections and the charge carriers it contains are significantly less numerous. The minute current that flows through the base is formed by electrons flowing from emitter to base and combining with existing holes. Most of the electrons flowing from the emitter pass by the base and flow through the collector, forming an I_C that is larger than I_E by the amount of the base current. In the body of a properly biased P-N-P transistor, the holes that travel from positive to negative carry current of the same magnitude as in the N-P-N counterpart.

The alpha (α) or forward current gain is the ratio of I_C to I_E, as follows:

$$\alpha = \frac{I_C}{I_E} \tag{2}$$

Since I_B is very small, this ratio is close to unity; the closer to unity the better the transistor.

1.2.3 Common Emitter Configuration

A transistor's configuration is termed common emitter when both supplies are connected to the emitter of the transistor. It is an extremely popular biasing method and is used frequently. Figure 1.11 shows a common emitter biasing circuit.

Equation (1) also shows the current relationship in this configuration and represents another way of biasing a transistor. The conduction is provided through forward biasing EB and reverse biasing the BC junction. A common emitter is also called *base biasing* model.

The two voltage sources can be replaced by a single power supply if the polarity of the new supply remains positive with respect to the transistor's emitter. I_B is smaller than I_C in this formation. The outputs are I_C and the collector/emitter

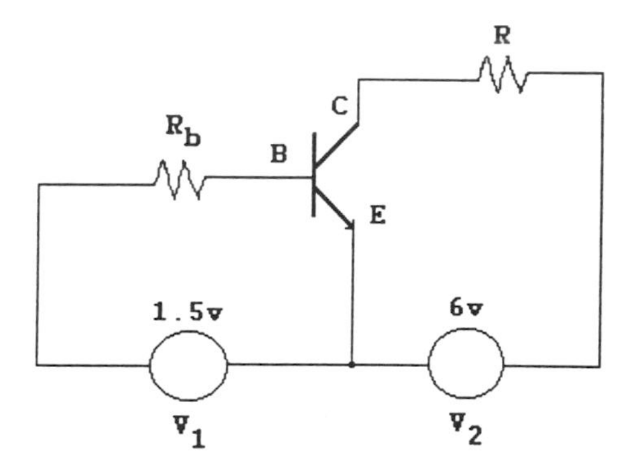

Figure 1.11 Common emitter biasing circuit.

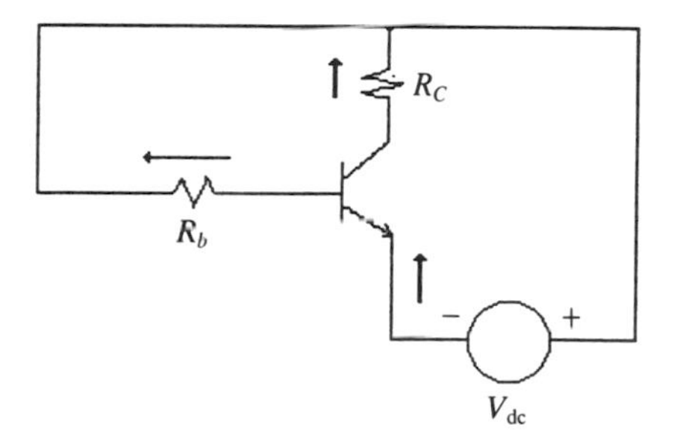

Figure 1.12 Single-supply common emitter schematic.

voltage (V_{CE}). The input is I_B. Figure 1.12 depicts a single supply common emitter connection.

The DC forward current gain, or β, is also called h_{FE}, and is the ratio of the I_C collector current to I_B, as follows:

$$\beta = \frac{I_C}{I_B} \tag{3}$$

Beta shows the control of I_B over the collector current and the above ratio is constant in any single transistor.

As mentioned previously, a transistor can function as a linear amplifier or as an on/off switch. A small signal at the base will produce a large I_C in phase with the input, if the transistor is in a common emitter form. There is no I_C without I_B. In the absence of I_B, the transistor is said to be in *cuttoff mode*; in this cutoff mode, it is the same as an open switch. If a base current is applied, the transistor conducts, becomes saturated, and simulates a closed switch.

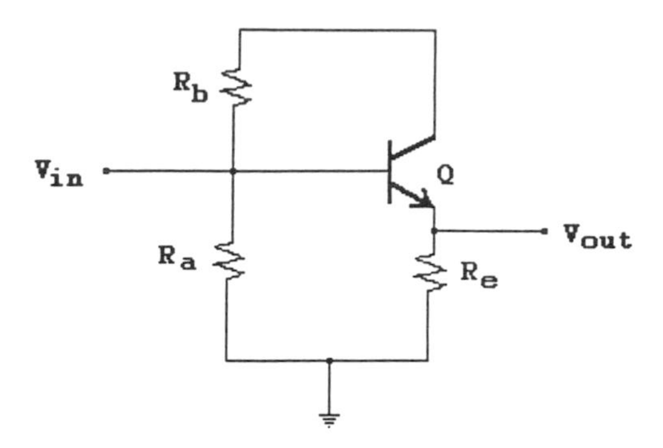

Figure 1.13 Common collector configuration.

1.2.4. Common Collector Configuration (Emitter Follower)

The common collector configuration functions over wide frequency ranges and is used in impedance stepping. The base is the input and its resistance is determined by the source and the load. Output resistance is low. The common collector connection has almost unity gain; its power gain is also β or h_{FE}. Figure 1.13 represents the common collector configuration.

1.2.5 Transistor Operation Modes

There are three modes of transistor operation: cutoff region, saturation region, and linear region. Each of these modes is described below.

1.2.5.1 Cutoff Region

This mode can be created in three ways:

1. Reverse biasing the BE region
2. Short circuiting the base to the emitter
3. Opening the base circuit (open base)

In this mode, the CB or the EB of a transistor is either not biased or reverse biased and the transistor does not conduct as there is a low flow of I_E and I_C. Contrary to theory, in the practical transistor, real cutoff does not exist because semiconductor materials contain enough imperfections to cause a limited flow of electrons, the *leakage current* mentioned earlier. This current can become a major problem when the transistor is used as a switch in elevated temperature conditions.

1.2.5.2 Saturation Region

Saturation occurs when the BE junction voltage exceeds breakdown voltage (>0.7 V in silicon), the normal V_{BE} for active mode. The transistor acts as a closed switch when saturated. With the EB and the CB junction forward biased, the transistor conductivity is substantial. I_B is elevated at the same time that I_C increases while V_{CC} stays the same.

The I_C becomes saturated because the voltage drop increases across the collector resistor R_C; this increased I_C causes V_{CB} to decrease. Increasing the I_B decreases the change rate of I_C and the transistor gain is low in this mode.

The switching occurs when the bipolar junction transistor (BJT) is driven through the active region from saturation to cutoff of vice versa very rapidly. The time for this transition to take place, although very small, is not defined since the transistor is neither in the On or Off position (binary states). How this undefined transition time can be minimized is of special concern to digital circuit designers. From the designer's point of view, there are three methods for achieving this goal.

1. Shorten the minority charge carrier lifetime in the base region of the transistor. This is directly related to the injection charges (dopes) in the base region and the collector current.
2. Increase the value of β.
3. Increase the value of I_B. This step is dependent on the circuit or elements connected to the transistor externally. A sudden drop in I_B turns the transistor to the Off state (cutoff region).

The off/on switch function as the change from cutoff mode to saturation mode occurs must be rapid to maintain switching stability in digital circuits. One of the most important characteristics of a transistor is its switching speed (transition between saturation and cutoff regions).

1.2.5.3 Linear (Active) Region

The linear (active) region is the area of the curve between saturation and cutoff (refer to Figure 1.8). All operating points are located there and great numbers of charge carriers travel from the positive emitter to the negative base. As mentioned previously, the EB is forward biased and the CB is reverse biased, and transistors amplify analog signals in the active region by the magnitude of β, which is the ratio of I_C to I_B. See Equation (3). The I_E and I_C collector currents create the amplification.

1.2.6 Bipolar Inverter Operation

Figure 1.14 provides an analysis of the operation of a bipolar inverter. In this mode, transistors operate as on/off switches.

The transistor is in the cutoff mode, not conducting, when the input voltage $(V_{in})=0$. With the EB forward biased,

$$V_{out}=V_{in}-V_{rl}$$

Therefore, the emitter is forward biased when $V_{in}=V_{CC}$ and the transistor conducts.

Also, with 0 V at the output, the transistor becomes saturated:

- Low input voltage results in high output.
- High input voltage results in low output.

An important characteristic of a transistor is the pulse propagated at the transistor output. *Propagation delay* is the time that a gate or a circuit requires to

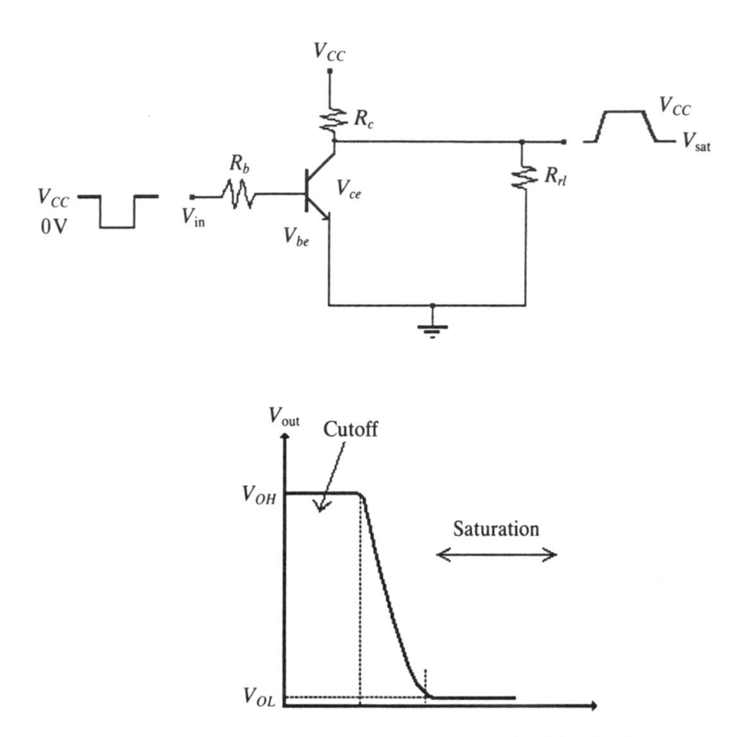

Figure 1.14 Voltage transfer characteristic of a bipolar inverter.

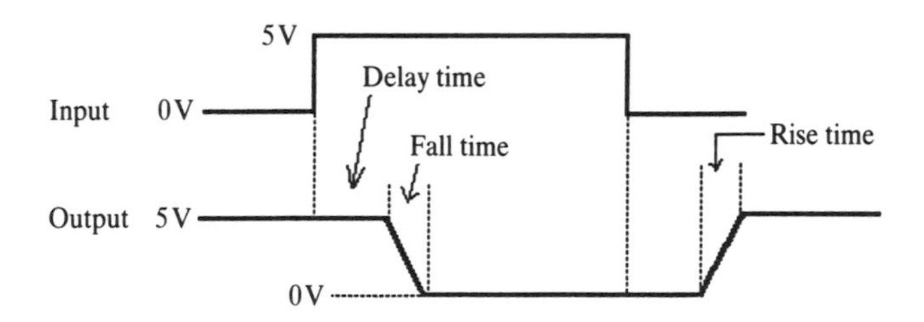

Figure 1.15 Input/output pulse curve.

respond to the applied signal at the input. Figure 1.15 illustrates some aspects of the output pulse in relation to time. This concept should be understood before continuing.

The following definitions are included for ease of understanding Figure 1.15.

- **Delay time.** The delay time is caused by the junction capacitance created at the emitter and collector of the depletion regions in the transistor, as was mentioned in Section 1.1. This is the time taken by the heavily saturated charge carriers to leave the base.

- **Fall time.** Fall time is the negative slope of the output pulse with respect to the input and is caused by the same junction capacitance. Fall time immediately follows delay time, then V_{in} changes to 0 V and the transistor stays in saturation mode with constant output voltage.
- **Rise time.** The output voltage returns to the supply voltage with a time delay called *rise time* after the transistor turns off. As in the previous cases, this time delay is also caused by the junction capacitance.
- **Transient time.** Transient time is the time required for minority charge carriers to pass through the base from collector to emitter. This affects the rise and fall times of the output pulse. A slight negative slope in the output pulse is caused by the junction capacitance. The output voltage is constant at this point and the transistor stays in saturation mode.
- **Recovery time.** The final stage of the output pulse is the recovery time. This is the time needed for the base/emitter voltage to change to 0 V. This time is not seen on the output pulse.

A pulse (amplitude 5.0 V) applied to the transistor input at time zero (t_0), will cause a change at the output at $t_0 + n$ (where n = delay time).

The bipolar family is divided into three groups:

- Transistor–transistor logic (TTL)
- Diode–transistor logic (DTL)
- Emitter-coupled logic (ECL)

Among these ECL has the fastest switching operation. These transistors are used in totem-pole output configuration for the lowest output impedance. This low output impedance slows the speed of the switching operation. A schematic of a two-input NAND gate totem pole configuration is shown later in Figure 3.16.

1.3 MOS Transistors

The metal oxide semiconductor field effect transistor (MOSFET) is another type of transistor that competes with the bipolar variety. This type of transistor is also used as a switch or amplifier. A MOSFET is a unipolar device in that it has only one type of charge carrier, holes in the P-type or electrons in the N-type. An important difference between the two types of devices is that bipolar transistors are current-controlled and MOSFETs are voltage-controlled.

Metal oxide semiconductors (MOS) transistors are made up of two P-N junctions joined by an MOS capacitor called a gate. Conduction between the source and the drain area is controlled by an electric field, which is created by applying a bias voltage at the gate between the junctions. Because of this electric field, these transistors are sometimes called field effect transistors (FET). The existence of the gate, which is completely isolated electrically from the source and the drain, is the reason for the terminology insulated gate FET (IGFET), or unipolar transistor. These transistors may be referred to as MOS, MOSFET, FET,

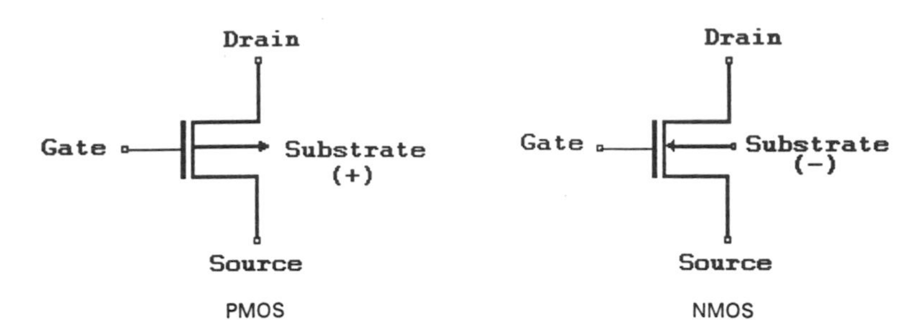

Figure 1.16 Symbols of MOS transistors.

MOST, and IGFET; there is no single universal name. A schematic symbol for a MOS transistor and its terminal designations are shown in Figure 1.16.

- The FET common source configuration is equivalent to the common emitter configuration in a bipolar transistor, and is used for voltage amplification.
- The FET common drain configuration is equivalent to the common collector configuration in a bipolar transistor.

MOS		*TTL*
Source	=	Emitter
Drain	=	Collector
Gate	=	Base

The Q point of an FET transistor is located on the DC load line and is chosen in the same manner as its bipolar counterpart. The drain/source voltage (V_{DS}) must always stay above pinch-off for proper biasing. These devices operate as constant-current sources at pinch-off and the magnitude of the current is determined by gate/source voltage (V_{GS}). The phenomenon of pinch-off voltage in MOS transistors is equivalent to saturation voltage in bipolars.

The source and drain of a MOS transistor are interchangeable, and can be correctly identified when a voltage is applied and polarities are known. In the MOS type, the drain is the heavily positive terminal, and in PMOS devices, the voltage polarities are the opposite of those in NMOS devices.

These transistors present extremely high input impedance, and are voltage-controlled devices. In contrast to bipolar transistors, in these the output current is controlled by the application of input voltage.

Two bias voltages are required to insure the operation of a MOS device. One voltage is applied between the source and drain (V_{DS}) such that the source becomes negative to the drain, and the other supply is connected between the gate and source, so that the P-gate is more negative than the N-gate. Ordinarily, V_{GS} is the only input voltage that controls the operation of the device. The current that flows out of the device to external load is the drain current (I_D).

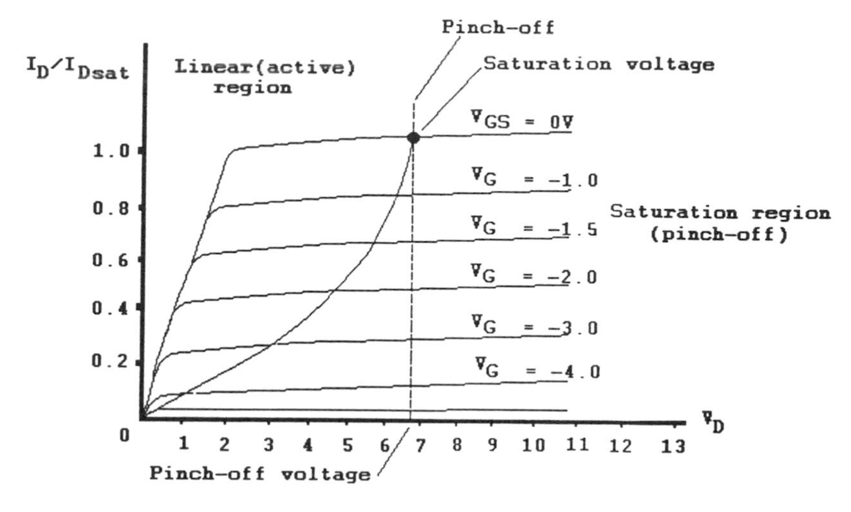

Figure 1.17 MOSFET drain current characteristic curve.

The cutoff voltage is produced when V_{GS} gradually increases until I_D becomes virtually zero. As in bipolar transistors, there are two types of junction FETs (JFETs), N-P-N or P-N-P type, divided into N-chanel or P-channel, and their operation, regardless of the polarity of the supply, is essentially the same. As V_{DS} increases, I_D also increases to a limited value. This increment of V_{DS} causes the depletion layer to expand to the point where it will pose resistance to I_D flow. I_D (at this point not increasing further) is called the pinch-off I_D. At this stage V_{DS} is called the pinch-off V_{DS}. Figure 1.17 shows a MOSFET drain current characteristic curve.

There is a quick way to find the polarity of JFET transistors using an ohmmeter. Connect the terminals of an ohmmeter between the drain and source: The ohmmeter shows a very low resistance if the ground side of the meter is connected to the source; otherwise, the resistance reading will be quite high. The resistance reading between the drain and the source is the same even if the leads are interchanged.

1.3.1 CMOS Inverter Operation

A complementary metal oxide semiconductor (CMOS) inverter has one N-channel and one P-channel transistor connected in a complementary-symmetric circuit. The input is common in either channel and the output is taken from the junction, where the source of one channel and the drain of another are connected. A simple CMOS inverter is depicted in Figure 1.18.

The operation of the CMOS inverter is as follows.

When input voltage = low (lower than threshold voltage)

Q_1 = OFF (high input impedance state) cutoff

Q_2 = ON (low impedance state) conducting

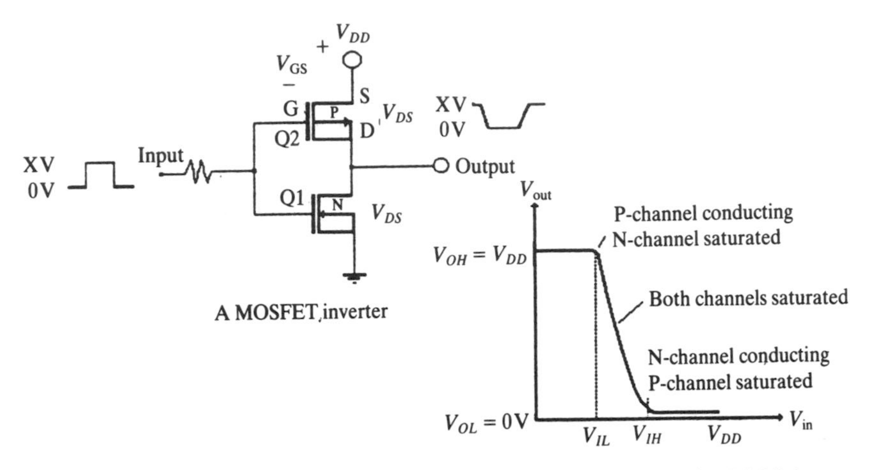

Figure 1.18 A CMOS inverter circuit, V and I transfer curve of a MOS inverter.

$$\text{Output} = V_{DD} = V_{OH}$$

$$V_{OL} = 0\,\text{V}$$

When input voltage = high ($\geqslant$ threshold voltage)

$$Q_1 = \text{ON (low input impedance) conducting}$$

$$Q_2 = \text{OFF (high input impedance) cutoff}$$

$$\text{Output} = \text{GND}$$

$$V_{GS} = 0\,\text{V}$$

1.3.2 CMOS vs. TTL Transistors

CMOS offers the following advantages over TTL transistors:

- **Current:** CMOS circuits draw considerably lower current in the microampere (μA) range, while an equivalent TTL device draws current in the milliampere (mA) range.
- **Supply range:** CMOS devices operate with a wide range of supply voltages. TTL devices operate only in specific supply ranges with the supply very near to V_{CC}.
- **Noise immunity:** CMOS devices have very low noise immunity because the state of the device changes only when the input reaches threshold voltage (usually $0.5V_{DD}$). Consequently, CMOS devices can tolerate more noise before it becomes an interference factor in their state. TTL transistors, on the other hand, change state when the input becomes close to supply voltage. The noise carried by the input can easily confuse the state of

the device. This factor demands the use of a decoupling capacitor on the unstable supply pin.

- **Size:** CMOS devices can be manufactured in far smaller sizes than bipolar devices. Consequently, many more CMOS devices can be packed into the same area.

1.3.3 Input Characteristics of a Bipolar Transistor

In general, all members of the TTL family have the same input characteristics and are tested in the same way. The sites for all semiconductor testing are the inputs and outputs of transistors and diodes.

When the input voltage of a TTL device rises above the threshold value the following happen:

- Switching takes place.
- The current at the base divides into the input transistor collector and emitter and the input current drops to its lowest value.
- Some of the input current leaks into the gate input.

Maximum input voltage is of great significance and leakage current should be kept as small as possible. If the voltage suddenly drops to zero, the chance of damage from leakage increases greatly. Two factors that can be used to minimize leakage are gold doping and the geometry of the transistor input pad in the fabrication process. A typical bipolar input characteristic curve is shown in Figure 1.19. I_{IH} represents input current at high output state, I_{IL} is input current at low output state, and I_I is maximum high level input current. In this curve, the current value depends on the transistor base resistor when there is no applied voltage at the input. The current value decreases as the input voltage increases until the input voltage reaches the threshold voltage.

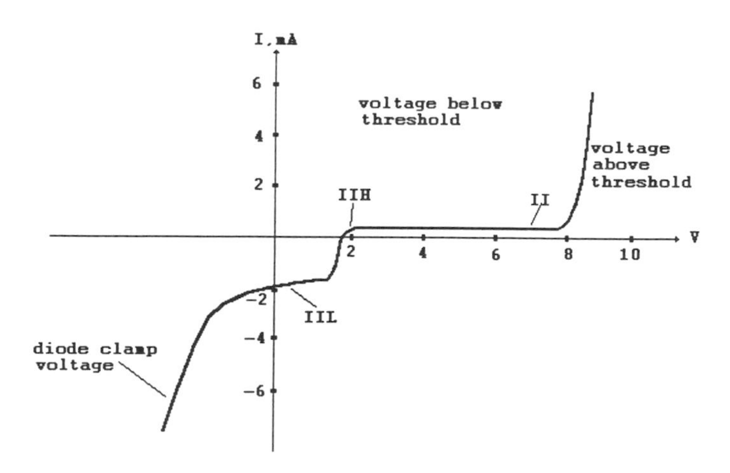

Figure 1.19 Input characteristic curve for a typical TTL gate.

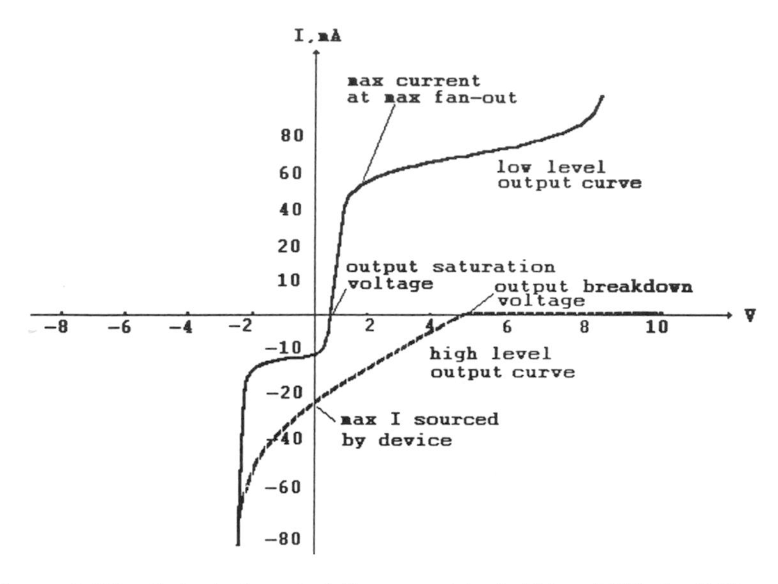

Figure 1.20 Output characteristic curve at logic "1" and "0" of a TTL gate.

1.3.4 Output Characteristic Curve of a Bipolar Transistor

The two curves in Figure 1.20 represent the high and low output properties of the bipolar gate. These two output levels represent the On and Off conditions of the switch and are designated V_{OL} and V_{OH} for low and high state of output, respectively.

- V_{OL} is the highest low level voltage the device output will reach for a given load when V_{DD} or V_{CC} is minimum.
- V_{OH} is the lowest high level voltage the device output will fall to for a given current load when V_{DD} or V_{CC} is minimum.

Low level Output Figure 1.20 depicts the normal output impedance of a transistor in saturation mode. The current rises sharply when high voltage is applied, which increases the chance of damage occurring because some current sinks the output when the applied voltage is below V_{OL}.

High-level Output When the applied voltage at the high-level output is greater than V_{OH}, a minor amount of current flows through the output until the input voltage reaches breakdown. At breakdown, the current through the output will rise sharply with small increments in input voltage. The flow of current is dependent on the ground level of the high output level. Set at ground level, the current flow is about 45 mA. Set below ground level, the flow increases and is independent of negative voltage.

References

1. Comer, J., *Modern Electronic Circuit Design*, Addison-Wesley, 1976.
2. Blakeslee, Thomas R., *Digital Design with Standard MSI and LSI*, Wiley, 1973.
3. Robinson, Vester, *Manual of Solid State Circuit Design and Troubleshooting*, Reston Publishing Co., 1975.
4. Healthkit/Zenith Educational System, *Semiconductor Devices, A Step by Step Introduction*, Prentice-Hall, 1983.
5. McCarthy, Oliver J., *MOS Devices and Circuit Design*, Wiley, 1982.
6. Millman, Jacob, and Christos C. Halkias, *Integrated Electronics: Analog and Digital Circuits and Systems*, McGraw-Hill, 1972.
7. Lenk, John D., *Handbook of Practical Electronic Circuits*, Prentice-Hall, 1975.
8. Marks, Myles H., *Basic Integrated Circuits*, TAB Book, 1986.
9. Boylestad, Myles H., and Luis Nashelsky, *Electronics—a Survey*, Prentice Hall, 1978.
10. Scarlett, J. A., *Transistor–Transistor Logic and its Interconnections, A Practical Guide to Microelectronic Circuits*, Van Nostrand Reinhold, 1970.
11. Cooke, M. J., *Semiconductor Devices*, Prentice-Hall, 1990.
12. Buchsbaum, Walter H., *Practical Electronic Reference Data*, 2d ed., Prentice-Hall, 1976.

CHAPTER 2

Integrated Circuit Test Basics

Introduction

This chapter provides a cornerstone for test engineering and should be comprehended fully before proceeding. While the actual design and fabrication of silicon wafers are not a concern of test engineers, they need to be familiar with the process and the types of tests that are performed on the devices. It is crucial to understand when each type of test is performed, why it is required, the test limits that apply, and the equipment that is used. Problems that are usually involved in wafer testing are pointed out and discussed from a test and product engineer's point of view. Although digital and analog circuits are entirely different, the procedure for testing them is basically the same. Once the philosophy of device testing is understood, the differences between the two types of circuits must be comprehended in order to design comprehensive test programs.

Topics

2.1. Designing ICs, including the use of computer-aided design (CAD), simulation, and production of masks.

2.2. Silicon wafer production, including the processes of deposition, photolithography, etching, and diffusion (or ion implantation) that form the die on the wafer:

- Testing of the ICs on the wafer by wafer sort, final test, and quality assurance (QA) testing
- Use of automated test equipment (ATE).

2.3. Test limits guardbanding, including

- Limits used
- Customer vs. manufacturer guardbanding, with examples

2.4. Wafer test hardware (with illustrations):

- Wafer sorting, use of probe cards, and vacuum chucks
- Test limits and test software

2.5. Bonding and packaging of the die following wafer testing:

- Scribe and break process (with illustrations)

2.6. Final test:

- Test limits guardbanding
- Military devices

2.7. QA test:

- The lot
- Writing test programs, including fundamentals and methodology of different time measurements—ACtesting, AC testing on a simple gate
- Quad NAND gate test procedure, an example of a manufacturer's data sheet, rise/fall, propagation delay, setup and hold times
- Generic algorithm for AC test software, with all necessary steps
- Algorithm for functional test
- DC or static parametric tests
- Programmable loads, description of Schottky diode

2.1 Designing Integrated Circuits

As technology moves forward, the various computer software tools used to design ICs increase in sophistication and capability. These tools are used by the designers to develop increasingly more complicated circuit designs.

Computer-aided design (CAD) computer software is used to design and test ICs. This software is very specialized and sophisticated. Previously designed cells are stored in the CAD design library. Each cell is specifically tailored in relation to the other cells to perform the intended function. The designer uses these cells to design each layer of the circuit.

CAD also has the capability of performing very lengthy and sophisticated tests on the circuit, referred to as *simulation.* Data representing the circuit with mathematical (transistor level) or Boolean (behavioral model) relations for each cell or block function is used to perform the theoretical simulation. Simulation convinces the designer of the functionality and total operability of the circuit being designed. It is important for test engineers to understand what can be tested by the simulation. At this stage, there is no die, wafer, or physical evidence of the actual circuit. Certain parameters, such as leakages and power supply rejection ratio (PSRR), cannot be simulated accurately. In such cases, historical data is used. Sometimes simulation hard copy can be very helpful to the test engineer for test debugging purposes.

After the circuit design is completed, it is converted to several masks, each containing the information of a layer of the circuit. The masks representing each layer are used to create the image of that layer. For example, there are individual masks for insulations, connections, bonding pads, and gate difinition. These masks are used to create the physical electronic circuit on silicon wafers. This circuit contains actual diodes, transistors, and other required electronic elements.

2.2 Silicon Wafer Production

A wafer is a slice of a processed silicon cylinder. The diameter of the silicon cylinder varies, usually between 3 and 8 inches, and is dependent on the technology and capability of the processing equipment. The surface of the wafer must be polished to perfection and inspected before the implantation process begins.

In the wafer process, the silicon wafer is exposed to oxygen. The reaction of silicon and oxygen forms a continuous layer of silicon dioxide (SiO_2 or glass) on the surface of the wafer. Silicon dioxide has excellent insulating characteristics that make it an ideal choice for use in integrated circuits. In this process a small percentage of the surface of the silicon substrate is consumed. The creation of each layer requires the following steps:

- **Deposition.** In deposition, the conducting layers, metal, epitaxial silicon (epi), metal silicides, and insulating material such as silicon dioxide and related materials, are deposited on the wafer. This process is critical because lack of uniformity in the thickness of deposited layers creates reliability problems. For example, irregular thickness of the metal layer causes nonuniform current density in the conducting route of the die, the higher current density being in the thinner layers.
- **Photolithography.** Photolithography is a technique used to transfer a desired pattern onto the surface of a processed silicon wafer. This is key in the IC integration process. Lithography can be achieved by X-ray, electron or ion beams. A photograph of each mask is reduced in size hundreds of times and then transformed into an actual electronic circuit. Photolithography is a process by which a channel or window is opened through the SiO_2 layer for injecting dopes (refer to Section 1.1 in Chapter 1). The resolution of the photolithography process determines the accuracy of the transfer of the pattern onto the wafer surface. Photoresists are light-sensitive chemical materials deposited on the surface of a previously polished wafer. The exposure of any specific pattern on the surface of such a wafer is determined through the masking operation.
- **Etching.** The deposition process is followed by etching, in which any material from the area of the wafer not covered in the photolithography process is removed/not removed (depending on the positive or negative photoresist being used). Etching can be done chemically by dissolution, physically by ion bombardment, or by a combination of both.
- **Diffusion.** In diffusion, the doping agent is introduced to the semiconductor substrate. A precondition for diffusion is the existence of a concentration gradient that facilitates the uniform distribution of atoms in the solid. Some common dopants are boron for P-type creation, and phosphorus, arsenic or antimony for N-type. Diffusion is carried out at high temperatures, approximately 1100°C; as the temperature drops below 700°C; the process ceases.
- **Ion implantation** is another technique for introducing the impurities (dopes) into the crystal substrate by ion bombardment. Ion implantation is carried out at lower temperatures and is more suitable for high-density microcircuit fabrication.

There is more control of the dopant profile in this process. There are two important factors in the ion implantation process. The first is the *dose* (that is the number of ions per cm^2), which determines the concentration of ions injected into the semiconductor, The second is the *energy* (in electron volts, eV), which determines the depth of penetration of ions inside the substrate.

By the end of the process, an image of the intended circuit is formed on the silicon wafer and is called a *die*. Dies are the actual, testable circuits. Each die on

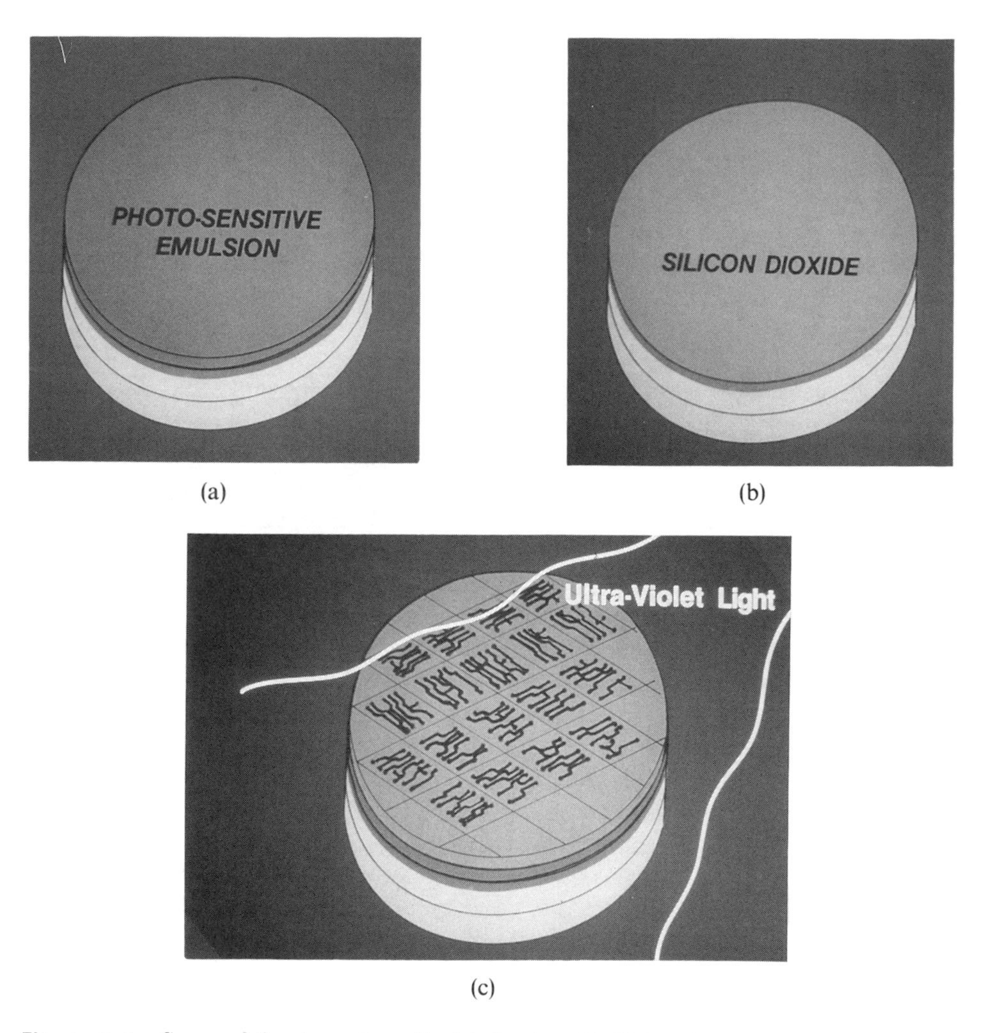

Figure 2.1 Some of the stages in wafer fabrication. (a) The surface of the wafer is now ready to absorb the image. (b) An example of oxidization of a silicon wafer. The wafer is exposed to oxygen under high-temperature conditions; the result is the formation of a thin layer of silicon dioxide, which is also called "glass". (c) Exposure of one layer of the circuit in photolithography. (Courtesy of Intel Corp.)

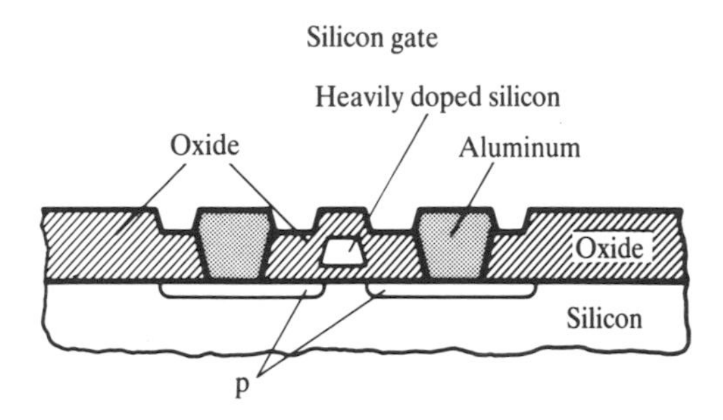

Figure 2.2 A typical cross-section of a MOS transistor implanted on the surface of a silicon wafer. (Courtesy of Intel Corp.)

a wafer is identical in size and function. The wafer contains many dies depending on the size of the wafer and of each die. Figures 2.1 through 2.4 illustrate the semiconductor fabrication process. Each wafer has a flat edge that is used for reference in the masking process and in wafer sorting.

The wafer, prepared to this point, is ready for wafer sorting. Every single die on the wafer must be electrically tested to insure the integrity of the circuit operation.

The role of the test engineer now comes into the scene. There are three classes of IC testing: wafer sort, final tests, and QA tests. Wafer sorting is performed on the wafer before the dies are cut. Final and QA tests are performed after the dies are packaged.

Testing dies on the wafer or packaged die is done by sophisticated application-specific computers termed automated test equipment (ATE). ATEs may be different in physical appearance and software design, but they all have the same purpose. A test strategy can help engineers and technicians regardless of the kind of ATE chosen as a tool. ATE manufacturers usually provide short-term classes for their customers in which the details of the system hardware and software are discussed. The information given in this book is for ATEs in general, since every ATE is designed differently and each operates under its own specific software. No ATE is complete enough to be responsive for testing every complicated device circuitry. A very complicated microcircuit can be designed within a few months, while the design of a responsive ATE may take years before it comes to the market. Consequently, ATE technology cannot keep pace with either the speed of operation or the complexity of circuitry developments.

The most important requirement, besides workable test software, is the test limits that are used for different parameters inside the test program. IC manufacturers have data books that explain the operation and limit measurement of every product they offer. Data books are the best, or the only, source of correct information on the ICs. Every parameter that is testable, along with its maximum

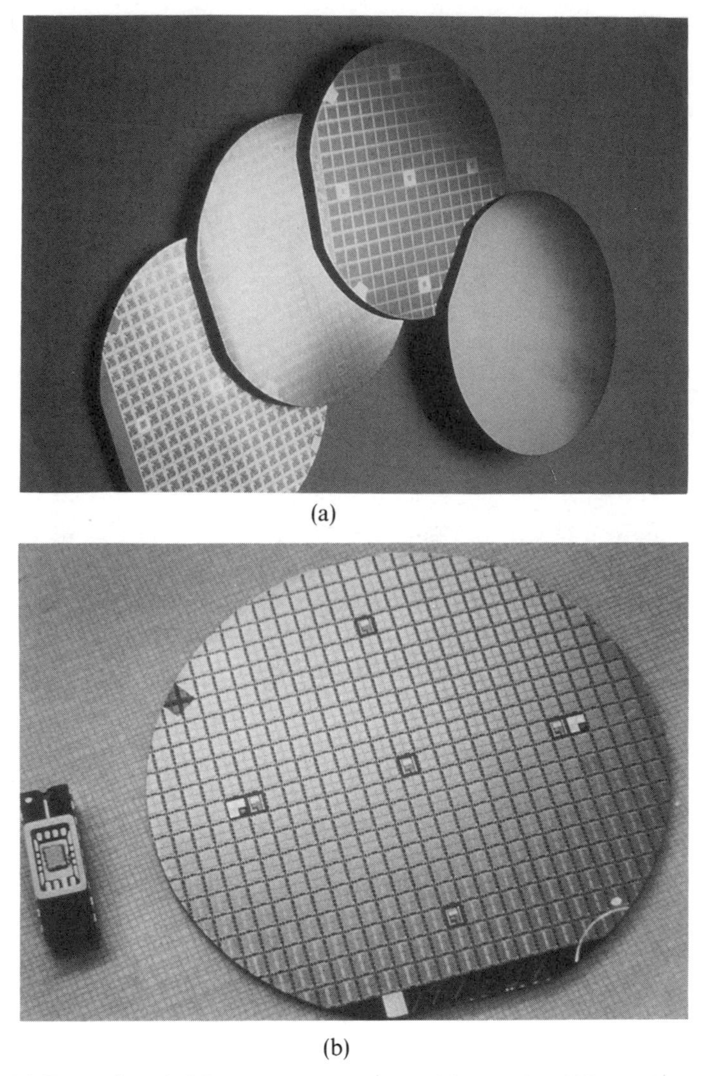

(a)

(b)

Figure 2.3 (a) Examples of different stages in photolithography; different layers of the designed circuit are implanted on the wafer at different stages. (b) A processed wafer containing dozens of similar dies; this wafer is now ready for test. (Courtesy of Intel Corp.)

and minimum limits, must be chosen accurately. Test limits guardbanding is one of the steps that must be undertaken in choosing the correct test limits in wafer sorting and final test.

2.3 Test Limits Guardbanding

In test engineering, guardbanding is applied to test limits. All parameter limits that are given in the manufacturers' data books are for quality assurance (QA) tests.

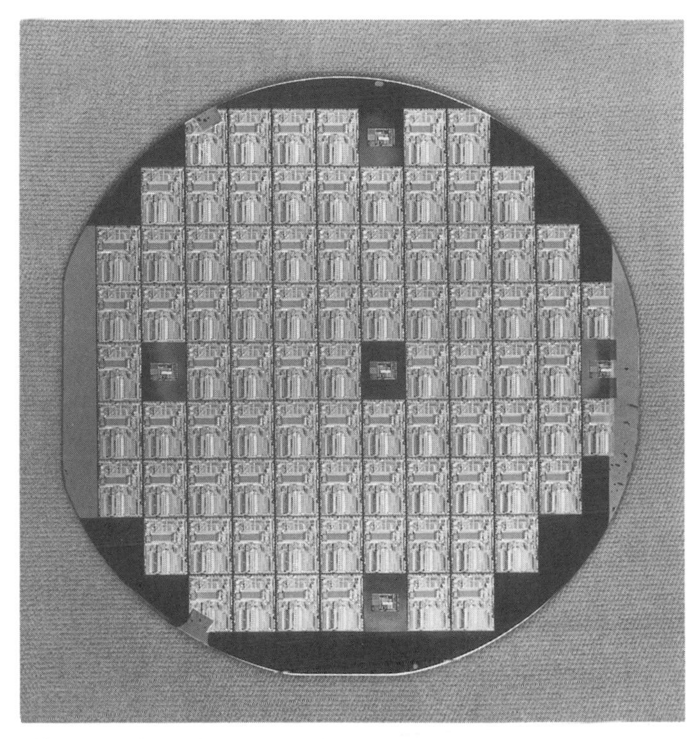

Figure 2.4 Close-up view of a processed wafer in which the layout of transistors individual dies is discernible. (Courtesy of Intel Corp.)

The maximum and minimum limits that are stated in the manufacturers' data books for any particular device are the limits guaranteed by the manufacturer based on the given conditions (mainly temperature). These are generous limits, meaning that the minimum and the maximum are much lower and higher, respectively, than the actual device performance. If the correct guardbanding limits are applied, any device that has passed wafer sort should not fail the final test unless there is physical damage to the device. The QA limits are not used in the production wafer sort and final testing for the following reasons.

1. The system tolerance (specifications) can vary under many environmental conditions such as temperature, humidity, and age. Most of the time, the ATE system specification given by the manufacturer is based on theoretical and mathematical evaluations, which are sometimes not very close to reality. The tolerance of forcing and measuring circuits inside ATEs varies depending on the factors mentioned earlier, which causes the marginal devices (those with some of their parameter limits very close to the maximum or minimum declared in the test software) to fail.
2. Most customers have their own in-house ATE that is used to verify the

integrity of the purchased devices before they are used. Customers usually test 100 percent of the purchased devices. The tolerance of the customer's ATE (same brand and type) may not be exactly the same as the chip manufacturer's ATEs for the reasons stated in (1). Every "identical" branded equipment will exhibit slight differences in measurement values under the same conditions, so no definitive "true" value can be assigned.—This discrepancy between measurements, based on the tools used, can create problems between the customer and manufacturer that can result in trashing of good devices. To prevent this kind of problem, manufacturers need to include worst-case tolerance variations of the test system, taking different environmental conditions into consideration.

The variations that affect voltage, current in either forcing or measuring mode, and time measurement must be included. The method of verifying these variations is called *machine guardbanding*. The guardbanded value is almost always in the form of $\pm X$, where X is either a real number or an integer.

Example If the guardband value of a particular V_{OH} test is $\pm 5\,\text{mV}$, this means that the measured V_{OH} should be decreased by 5 mV for the maximum and increased by 5 mV for the minimum limit. This $\pm 5\,\text{mV}$ limit must be accurate enough to cover the ATE tolerance variations and is called the *one machine guardband.*

Test limits guardbanding has its own philosophies and procedures. No universal set rule exists because uncontrollable factors can affect circuit operation in the ATE, in the interfacing hardware, or in the device under test itself. By now the difficulties may be obvious; once the above factors are identified, then a workable formula for guardbanding limits is needed. Thus, an accurate and exact guardbanded limit does not exist, and guardband limits are relative. Different IC manufacturers may have different guardbanding rules for different families of products.

Figure 2.5 shows the relationship between machine guardbanded limits of the three test classes.

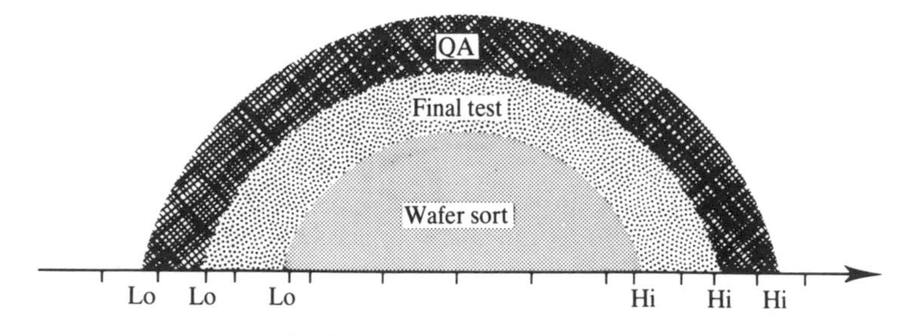

Figure 2.5 A graphical representation of guardbanding relationships.

2.4 Wafer Test Hardware

Testing each individual die on the wafer is called *wafer sorting*. The equipment used for this purpose is called a *prober*. The wafer is set on a vacuum chuck that is totally software controlled during testing.

Probe card hardware is needed to connect the force and measurement units of the ATE to the respective pads on each die. Probe cards may have different sizes or shapes for different dies. Every probe card is designed and fabricated for only one type of die. The number of needles and the spaces between needles on the probe card is exactly the same as the number of pads and the spacing between each pad on every die.

The chuck holding the wafer must now be adjusted with the probe card so that every needle of the probe card touches its respective die pad. This adjustment procedure requires experience and skill. The adjustment must be done very accurately so that every die on each row and column of the wafer is uniformly touched by needles throughout the wafer testing. In modern wafer sorting equipment this process is performed automatically. The die size and die pad measurements are given to the system to facilitate the automated adjustment process through a few reference points on different parts of a wafer.

A ribbon cable usually connects the probe card to the test equipment. Each wire strand in the ribbon cable is connected to an individual needle of the probe card and from there to the die pads, which are the input, output, ground, power supply, and other pins of the die circuit. In other words, one end of the wire in the ribbon cable touches the pads on the die and the other end is in touch with the force and measurement modules of the ATE.

Figure 2.6 shows a typical probe card with the ribbon cable connection and

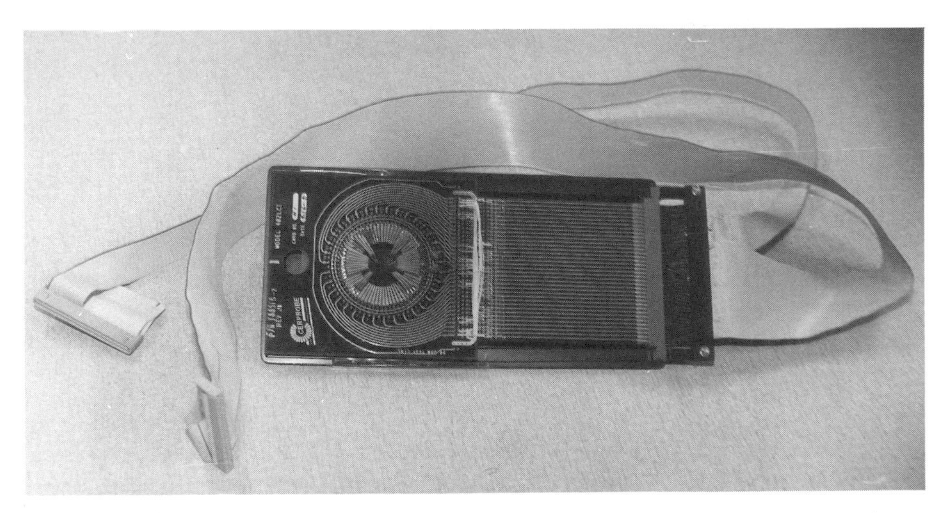

Figure 2.6 A typical old-fashioned probe card hardware and its ribbon cable connection. This hardware is essential equipment for wafer testing. (Courtesy of CerProbe Corp.)

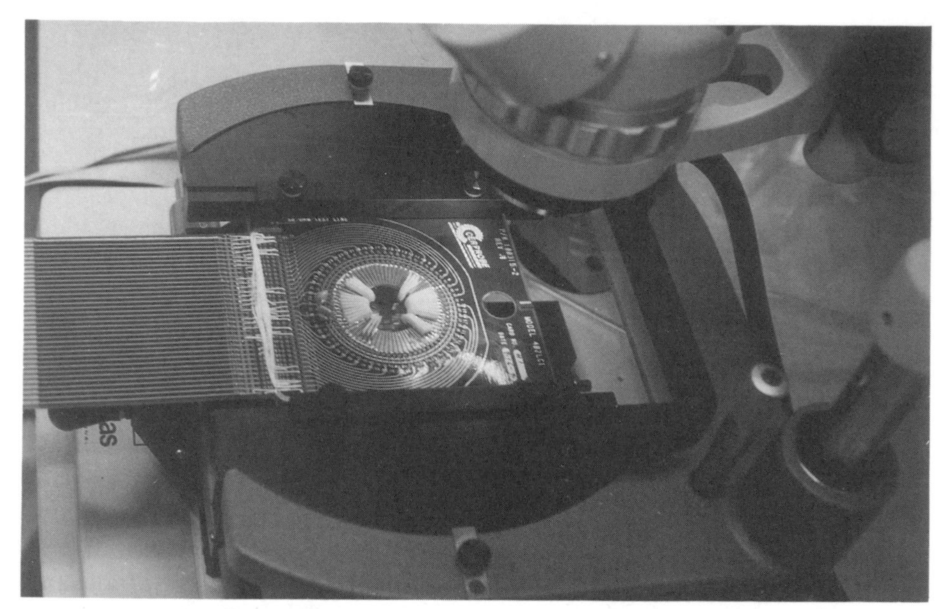

Figure 2.7 The probe card of Figure 2.6, now mounted on the special automated machine designed for testing dies on wafers. (Courtesy of CerProbe Corp.)

Figure 2.8 A close-up view of a die undergoing electrical testing in wafer sorting stage. The inker is adjusted on and above the center of the die. (Courtesy of Intel Corp.)

Figure 2.7 shows the probe card assembly mounted on the wafer prober. Figure 2.8 shows a close-up of a wafer undergoing test. Note the location of the probe needles and also the inker at the center of and above the die.

All of the DC and functional tests in the software program are applied to the die in the short time that the probe card is in contact with it. DC tests involve current and voltage measurements. In the case of digital devices, functional tests verify the operation of the device under test by the application of the related truth table sequence to the input and observing the predefined response at the output of the device. The actual level of the output pulse is usually not the criterion in functional testing; the output level is compared with predefined "high" and "low" voltage levels assigned by the test engineer to the comparators inside the ATE. This is the basis for terming the test a *go/no go* or *pass/fail* type.

At the conclusion of the test program, the power-down routine cuts off the probe needles from force and sense capability with the ATE. At this point the chuck moves horizontally to the next die, where the needles again sit on the next die pads and the whole sequence of tests restarts.

The power-down subroutine ending the software, or failure detection, terminates the test program with power cutoff to the probe card. Each failed die, regardless of the reason, is considered a bad die and distinguished from good dies by a small, dark ink spot placed on the surface. The operation of the inker is also controlled by the software program. In modern wafer sorters the inking issue is obviated by the wafer mapping capability of ATEs. Dies that have failed are recognized by the assigned bin number they receive during test.

The test limits that are used in water sorting are *two machine guardbanded* QA test limits. In other words, the limits (QA) that are declared in data books for any particular device parameter will be guardbanded twice (the maximum and minimum values of the QA limits are reduced and increased, respectively, by two machine guardband values) for wafer sorting. This will insure that marginal devices are caught at this stage before going through the costly packaging process. Marginal devices are those that pass one or a few tests with measured limits very near to QA. These devices will obviously not be able to pass the final and QA tests for reasons such as package stress on the die, moisture trapped in the package, or minor temperature changes.

Probe resistance can sometimes create problems in very low voltage measurements. The best remedy for this is to replace the particular needle that touches the die pad with another needle made of lower-resistance conductor.

Test software must be written so that, when a die fails, the next software code will not be executed. When a failure occurs, the program should immediately jump to the power-down subroutine and indicate the bin number of the last-executed test. This saves considerable test time. Every test module in the structure of the test software program is assigned a bin number. It can be replaced by other messages set by the manufacturer as part of the test software rules. The bin that indicates a good device does not belong to any particular test module. This occurs when all of the test modules in the test program are executed and no failure occurs. Any test that fails calls the power-down subroutine and the test flow stops at that point. In this case, the bin number reflects the failed test.

Since the length of the cable or ribbon connector creates some transmission line effects (capacitance, inductance, and impedance) between the circuit on the die and the measurement units inside the ATE, certain test limitations arise. These limitations usually prohibit dynamic tests (time measurement), especially rise and fall times in wafer sorting. Rise and fall times will be discussed later in this chapter.

In conclusion, the die goes through all functional and DC tests, but *not* all dynamic tests. It is obvious that if the distance between the die and force/measurement units of the ATE were considerably reduced, and the stray capacitance or inductance of the ribbon cable were not adverse factors, then all AC measurement could be done at this stage. Any AC measurement declared in this book refers to time measurement, also called *dynamic test.*

2.5 Bonding and Packaging

The tested wafer will be glued onto a nylon sheet before going through the scribe and break process. In this process, a fine diamond saw, or laser scriber, cuts the wafer with horizontal and vertical lines along the narrow spaces between the dies. The cut is not through the entire thickness of the wafer, and a roller then exerts uniform pressure on the surface of the wafer, causing it to break along the scribed lines, separating rectangular dies from each other. Inked dies are separated and trashed, and good dies are saved individually and glued onto a supporting base.

The next process is packaging, starting with bonding. In this step, very fine gold wires connect each bonding pad on the die to an appropriate pin, which will be the external connection to the die. Conducting glue or heat-generated impact are used to make these connections.

The final step is the completion of packaging, where the die is covered and hermetically sealed. All the above processes are done automatically and with amazing speed and accuracy. The packaged die is now ready for the final and quality assurance tests. Figure 2.9 illustrates the die assembly; Figure 2.10 illustrates the steps in the bonding and packaging process. Figure 2.11 shows a closeup of a die showing external connections to the pads on the die, known as "bonding pads." A bonding pad is shown in greater detail in Figure 2.12.

2.6 Final Test

In the final test, devices are 100 percent tested. All tests that were not possible in wafer sorting are now included.

The ribbon connector used in wafer sorting is not needed. Devices under test (DUTs) are tested either manually (for experimental and data collection purposes) in the socket mounted on the loadboard (the interface board between the DUT and the pin electronics modules on the ATE head), or by the aid of an automatic handler in production testing. In either case, DUTs are in close contact with the ATE and there is no transmission line effect of the ribbon cable.

The test limits in the final test program are somewhat looser (wider margin between minimum and maximum limits) than in wafer sorting (refer to Figure 2.5). *The final test limits are one machine guardbanded QA test limits.*

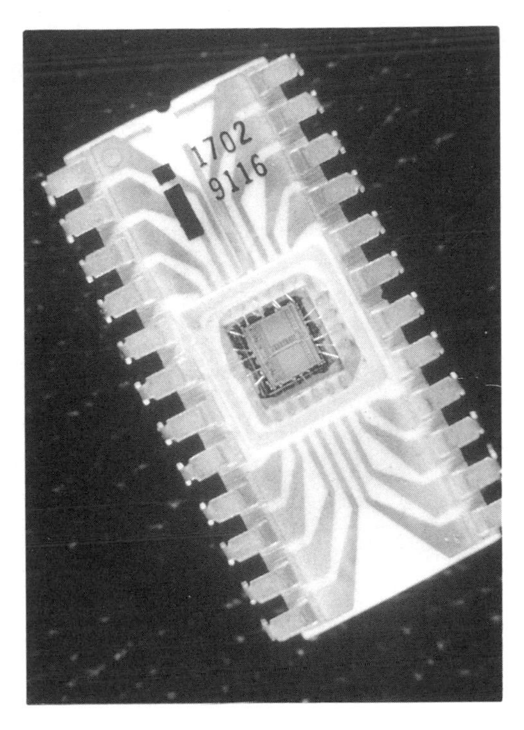

Figure 2.9 A tested die is placed inside the package. Bonding wires originate from the pads on the die (see Figure 2.12) and end at the package pins. (Courtesy of Intel Corp.)

Tubes of untested devices are emptied into the handler channels to be tested individually. Bad devices will be separated on the basis of the assigned bin. Two different types of automatic handlers widely used in final and QA tests are shown in Figure 2.13.

The test program can be written so that the good devices can be sorted according to quality. High-quality devices that have passed all tests with wide limit margins between high and low may be classified among the *military* devices. Quantification of military devices involves more constraints such as exposure of the device to very high or very low temperatures for a specified time duration as well as other electrical, electromagnetic, physical, and thermal shocks. The limits within which military devices pass tests are much tighter (their minimum limit is higher and their maximum limit is lower by specified amounts than for commercial devices).

2.7 Quality Assurance (QA) Test

QA testing is the next step a device must undergo before entering the application field. In this process, only selected devices in a lot undergo QA testing, depending on the integrity of the software program and the previous history of

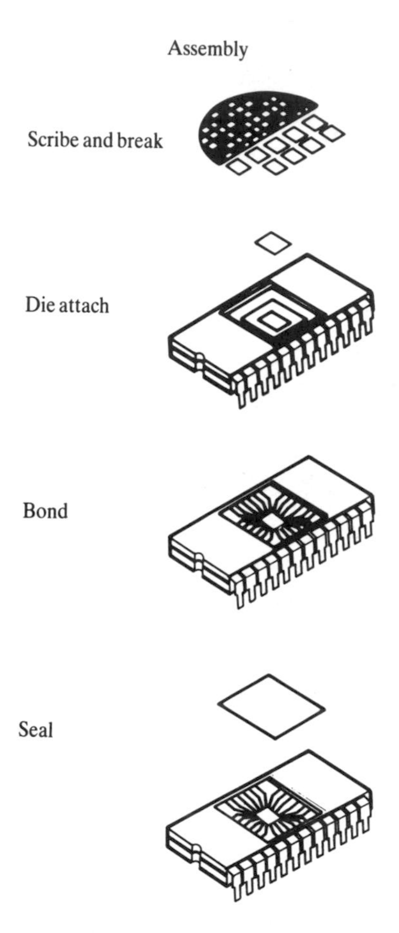

Figure 2.10 Steps in the bonding and packaging process. (Courtesy of Intel Corp.)

the device. A *lot* comprises all the individual wafers of the same device processed (manufactured) simultaneously. The QA test uses the same program as wafer sorting and final test. The difference is that QA includes all tests. The limits in QA programs are the same limits as declared in the device data book. Every device type has its own dedicated test software program. Not only does every test program contain all the parameters declared in the manufacturer's data sheet, but each parameter is used in three different test classifications (wafer sort, final test, and QA) where each has its own specific maximum and minimum limits based on the appropriate limit guardband.

A test program can be written in many different ways based on set rules or on the test developer's preference. Any test program should follow the universal rule of software writing—it must be understandable to other test engineers. In high-level structured computer languages, conditional words such as 'if,' 'then,' and 'else' enable software writers to create loops within a loop or procedures

Figure 2.11 A close-up of a die showing external connections to the bonding pads. (Courtesy of Intel Corp.)

within a procedure. This capability facilitates different categories and classifications in any single test block within the test program.

Every new program must be debugged if any test is to be 100 per cent repeatable. Test program debugging is usually achieved by reading a similar value for every test in every pass on a known good sample device. This criterion is crucial for highly sensitive parameters such as offset voltage in operational amplifiers, and current leakages. Noise is the principal factor in test repeatability.

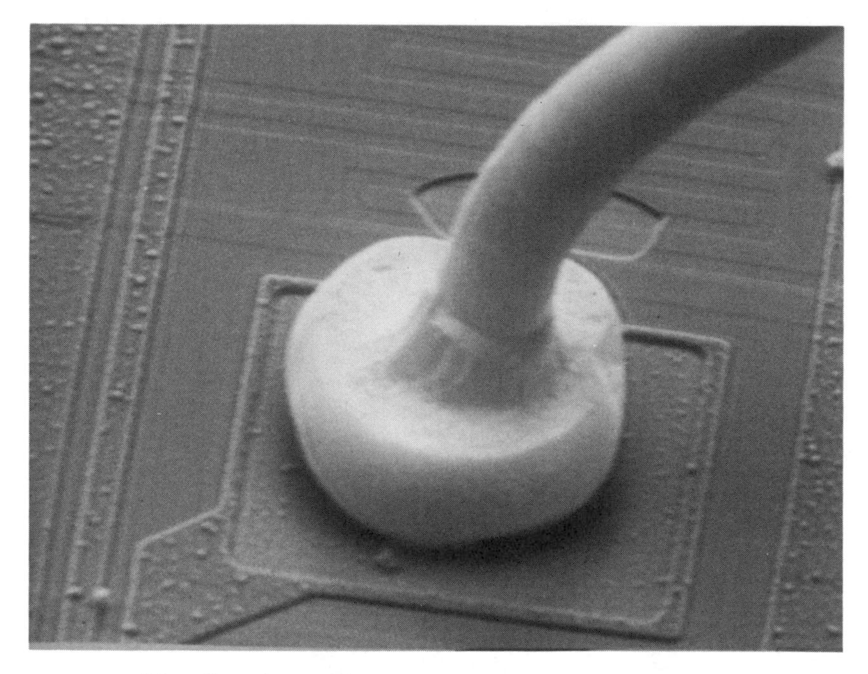

Figure 2.12 A bonding wire attached to a bonding pad of the die. (Courtesy of Intel Corp.)

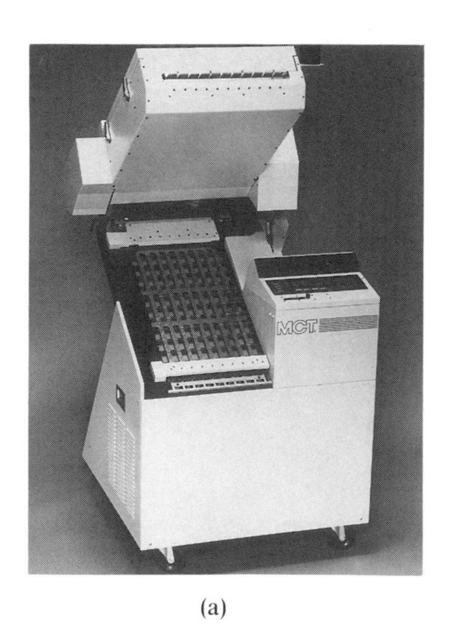

(a)

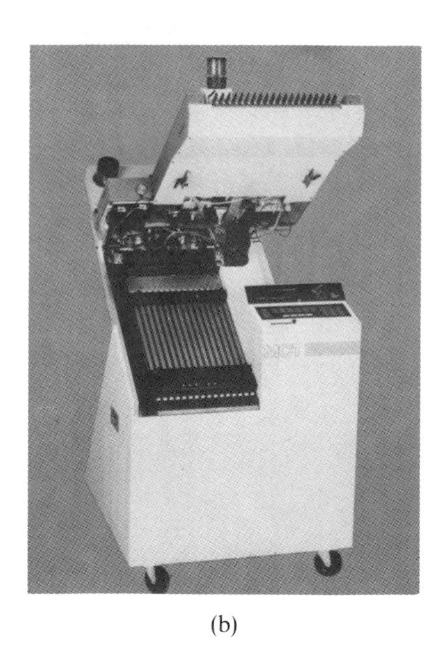

(b)

Figure 2.13 Widely used automatic test equipment for testing of packaged dies: (a) automatic handler for testing SMD type packages; (b) automatic handler for testing DIP type packages. (Courtesy of Micro Component Technology, Inc.)

A good test engineer develops a noiseless test for both software and hardware and minimizes the test time for cost-savings purposes. The strategy for test time reduction is avoidance of unnecessary waiting or delay time in the body of the test and reducing the necessary delay time so long as the integrity of each test remains intact.

The goal is to provide customers with the most reliable product at a minimum cost to the manufacturer. The effect of temperature on voltage and current is the fundamental consideration of semiconductor physics here. The effect of heat is not the same on MOS and TTL devices and this is reflected in the manufacturers' data books. As stated previously, limits under temperature extremes are called *military limits* and devices that have passed all those test limits are termed military devices. Military devices usually have different identification from commercial devices.

2.7.1 AC Parametric Test

Any test involving time measurement is usually called an AC or *dynamic test*. The most common tests are propagation delay and rise and fall times. Understanding the procedure of AC testing, as with other tests, is vital in designing workable test software on an ATE.

Propagation delay is the measurement of time between the application of an input signal and the related output response. This verifies the speed of the device operation, and is usually measured from 50 per cent of the rising or falling edge of the input pulse to 50 percent of the rising or falling edge of the output pulse.

Rise/fall testing is the same in essence. In this test, the slope of the edge of the output pulse (usually from 10 to 90 percent) is measured. Rise and fall times are dependent on the numbers and kinds of components forming the device internal circuitry. In some devices, the output pulse may not have sharp slopes. If the same output pulse passes through many such stages, its high amplitude level may not be sufficient to trigger further stages. The rise time is seriously affected by capacitance (refer to Section 2.4).

AC tests usually require some load at the device output. A certain amount of current is drawn from the device output during the device functional operation to verify its fan-out drive capability. Other AC parameters are pulse width, setup time, hold time, slew rate, etc. Detailed procedures for all AC tests are given in Chapter 3.

2.7.2 AC Testing General Information

AC testing includes several methods in various ATEs. Some of the methods are as follows.

- **Dynamic functional test.** This test uses timing parameters such as clock and data strobe at their worst-case values. It is a quick test but may be inaccurate due to strobe placement. Passing the functional test is considered to be equivalent to passing AC testing unless there is doubt about AC measurement or device malfunction. The test can be used in combination with the following sliding strobe test.

- **Sliding strobe test.** This is a dynamic functional test. Strobing is repeated at many locations until one setting passes and all other fail. This test is useful in finding the transition time at the output but has the disadvantages described in the previous method. It can be used in combination with the above dynamic functional test.
- **One-shot measurement.** This method uses special hardware for a one-shot (one-sample) measurement of the AC event. It is fast, but not accurate.
- **Real-time sampling method.** A specific pulse train is sampled and measured using an input signal provided by dynamic functional testing. This method is fast and fairly accurate since repeated events are tested for each AC parameter being sampled and measured.

The important factors in AC measurement are temperature, voltage swing, measurement point, and device loading. Figure 2.14 shows a sample data sheet for a quad NAND gate provided by National Semiconductor Corporation.

Test limits and corresponding power supply ranges for commercial (0° to +70°C for DM74...) devices are given at room temperature and for military (−55° to +125°C for DM54..) parts at extreme temperature.

Using software commands, a number of load configurations (called programmable loads under I_{OH} and I_{OL}) can be applied to the DUT output. Current equivalent to each type of load can be sourced (I_{OH}) to or sinked (I_{OL}) from the DUT output.

Examples of some external loads are given in Figure 2.15.

Automated test equipment provides a set of wave formats. A wave format is chosen for testing based on how easy it is to manipulate given the functional characteristics of the DUT.

Recommendations

1. Use functional patterns (based on the device truth table or timing diagram) format for the constant data through the test period.
2. Use NRZ format for one transition per test period.
3. Use RZ format for either positive or constant low pulse.
4. Use RT0 format for negative or constant high pulse.

Figures 2.16 and 2.17 depict some of the common wave formats available in ATEs.

High-performance and very accurate timing generator (t_g) circuits are employed in all semiconductor automated test equipment. Most ATEs, besides having a single period generator circuit, have many other general-purpose type circuits that are coherently connected to the period generator. The cycle time and its component (see Figure 2.20), as well as other aspects of time measurement, are controlled by the period generator within the ATE.

2.7.3 AC Test Format on a Simple Gate

To use various methods and meet the requirements of ATE equipment efficiently, it is necessary to understand the hardware and software of each system. The timing diagrams for different waveforms (based on the gate truth table or timing diagram) required to do functional testing on a selected NAND gate as shown in Figure 2.14 are shown in Figures 2.18 and 2.19. These are required for AC and functional test operation of the selected device in our example.

In the step illustrated, either input can be kept at logic '1' (V_{IH} level) while changing the pulse level from V_{IL} to V_{IH} at the other input. Obviously, output 3 in Figure 2.18 stays low all the time, because there is no transition at either of the two inputs. This input pulse does not facilitate proper conditions for AC measurement because it does not prove the functionality of the gate.

Figures 2.19 and 2.23 show the NRZ format chosen for both inputs.

AC measurement is possible with either of the two conditions listed here. In Figure 2.19, input 1 alternates between high and low while input 2 remains high. The output switches alternately to provide suitable conditions for AC measurement.

It is also possible to do the functional test on the NAND gate using the RZ format with input 1 alternating high and low and NRZ format held high at input 2. The output switches accordingly provide the correct operation for functional test. Either of the above conditions provides AC measurement conditions.

Other components of the pulse such as cycle time and its components can be controlled by the test software.

The following two timing selections must be made before any time measurements are performed:

1. The cycle time determines how fast operands (functional patterns, the sequence of the device's truth table) get to the DUT.
2. The start and stop times define the leading and trailing edges of the clock pulse.

These timing selections are dependent on two factors: The first is the specified frequency of the DUT as well as the timing generator capability of the ATE. The second is that the time between the start and stop might be larger than the test limit.

Figure 2.20 reveals the following information:

- Set t_g cycle 200.00 ns (specifies the complete cycle time).
- Start at 35.00 ns (specifies the leading edge of the pulse).
- Stop at 125.00 ns (specifies the trailing edge of the pulse).
- Strobe at 150.00 ns (specifies the point where strobe edge occurs).

The following steps are required to measure the propagation delay and rise and fall time for the NAND gate in our example.

1. Power supply voltage.
2. Input voltage levels including V_{IH} and V_{IL}.

DM54ALS00A/DM74ALS00A Quad 2-Input NAND Gate

General Description

This device contains four independent gates, each of which performs the logic NAND function.

Features

- Switching specifications at 50 pF
- Switching specifications guaranteed over full temperature and V_{CC} range
- Advanced oxide-isolated, ion-implanted Schottky TTL process
- Functionally and pin for pin compatible with Schottky and low power Schottky TTL counterpart
- Improved AC performance over Schottky and low power Schottky counterparts

Connection Diagram

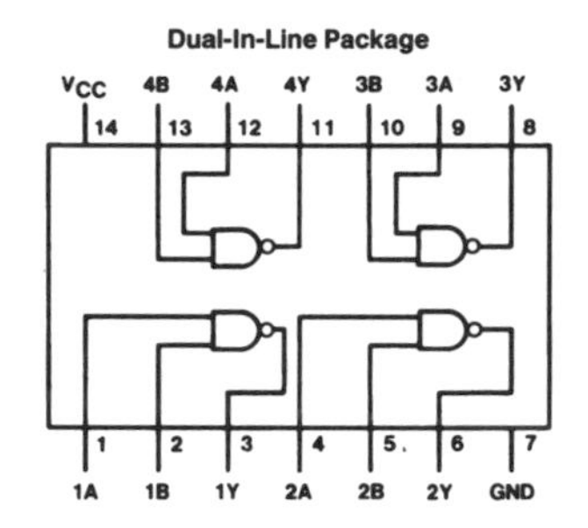

Order Number DM54ALS00AJ, DM74ALS00AM, DM74ALS00AN or DM74ALS00ASJ
See NS Package Number J14A, M14A, M14D or N14A

Function Table

$Y = \overline{AB}$

Inputs		Output
A	B	Y
L	L	H
L	H	H
H	L	H
H	H	L

H = High Logic Level
L = Low Logic Level

Figure 2.14 Sample data sheets for National Semiconductor Corporation Quad NAND gate. (Reprinted courtesy of National Semiconductor Corp.)

Absolute Maximum Ratings

If Military/Aerospace specified devices are required, please contact the National Semiconductor Sales Office/Distributors for availability and specifications.

Supply Voltage	7V
Input Voltage	7V
Operating Free Air Temperature Range	
DM54ALS	−55°C to +125°C
DM74ALS	0°C to +70°C
Storage Temperature Range	−65°C to +150°C
Typical θ_{JA}	
N Package	86.5°C/W
M Package	116.0°C/W

Note: *The "Absolute Maximum Ratings" are those values beyond which the safety of the device cannot be guaranteed. The device should not be operated at these limits. The parametric values defined in the "Electrical Characteristics" table are not guaranteed at the absolute maximum ratings. The "Recommended Operating Conditions" table will define the conditions for actual device operation.*

Recommended Operating Conditions

Symbol	Parameter	DM54ALS00A			DM74ALS00A			Units
		Min	Nom	Max	Min	Nom	Max	
V_{CC}	Supply Voltage	4.5	5	5.5	4.5	5	5.5	V
V_{IH}	High Level Input Voltage	2			2			V
V_{IL}	Low Level Input Voltage			0.7			0.8	V
I_{OH}	High Level Output Current			−0.4			−0.4	mA
I_{OL}	Low Level Output Current			4			8	mA
T_A	Free Air Operating Temperature	−55		125	0		70	°C

Electrical Characteristics

over recommended operating free air temperature range. All typical values are measured at V_{CC} = 5V, T_A = 25°C.

Symbol	Parameter	Conditions		Min	Typ	Max	Units
V_{IK}	Input Clamp Voltage	V_{CC} = 4.5V, I_I = −18 mA				−1.5	V
V_{OH}	High Level Output Voltage	I_{OH} = −0.4 mA V_{CC} = 4.5V to 5.5V		V_{CC} − 2			V
V_{OL}	Low Level Output Voltage	V_{CC} = 4.5V	54/74ALS I_{OL} = 4 mA		0.25	0.4	V
			74ALS I_{OL} = 8 mA		0.35	0.5	V
I_I	Input Current at Max Input Voltage	V_{CC} = 5.5V, V_{IH} = 7V				0.1	mA
I_{IH}	High Level Input Current	V_{CC} = 5.5V, V_{IH} = 2.7V				20	μA
I_{IL}	Low Level Input Current	V_{CC} = 5.5V, V_{IL} = 0.4V				−0.1	mA
I_O	Output Drive Current	V_{CC} = 5.5V	V_O = 2.25V	−30		−112	mA
I_{CC}	Supply Current	V_{CC} = 5.5V	Outputs High		0.43	0.85	mA
			Outputs Low		1.62	3	mA

Switching Characteristics over recommended operating free air temperature range (Note 1)

Symbol	Parameter	Conditions	DM54ALS00A		DM74ALS00A		Units
			Min	Max	Min	Max	
t_{PLH}	Propagation Delay Time Low to High Level Output	V_{CC} = 4.5V to 5.5V R_L = 500Ω C_L = 50 pF	3	15	3	11	ns
t_{PHL}	Propagation Delay Time High to Low Level Output		2	9	2	8	ns

Note 1: See Section 1 for test waveforms and output load.

Figure 2.14—*continued*

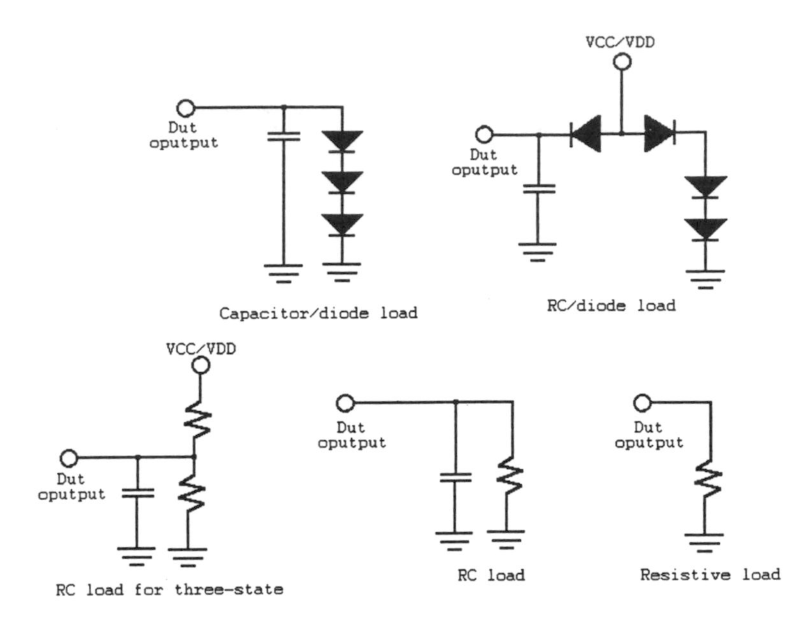

Figure 2.15 Examples of external load configurations.

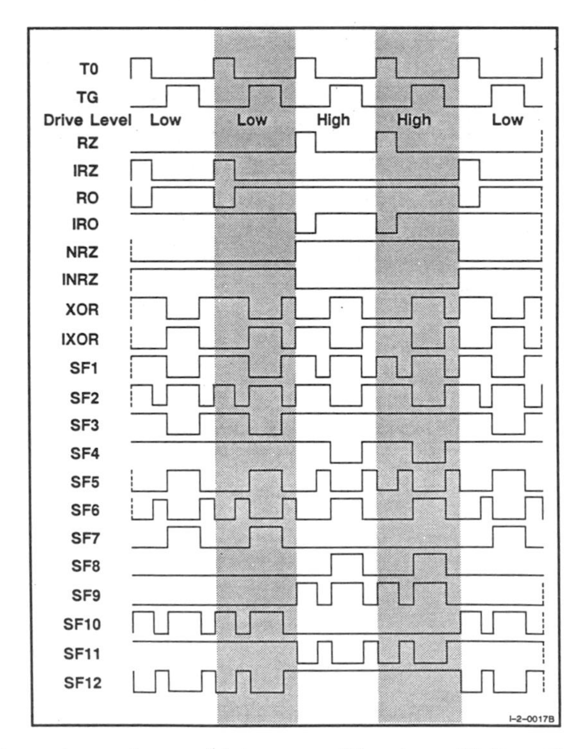

Figure 2.16 Format waveforms, driving at t_0. (Courtesy of Micro Component Technology, Inc.)

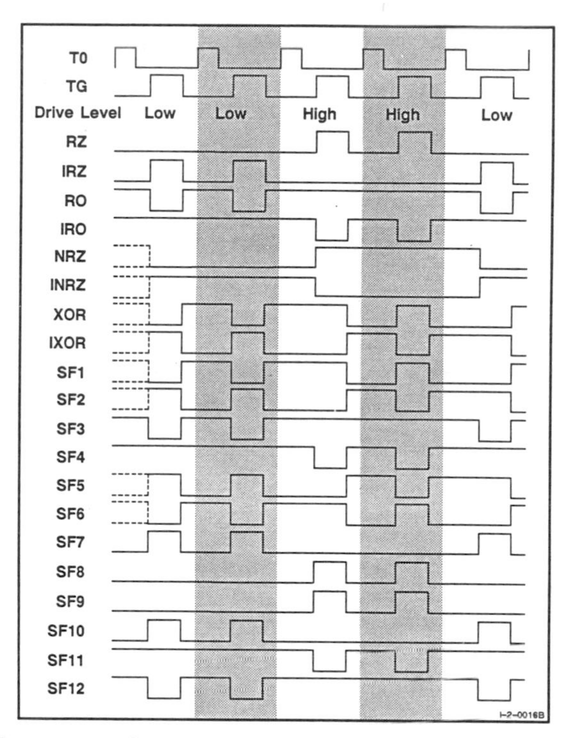

Figure 2.17 Format waveforms, driving at t_g. (Courtesy of Micro Component Technology, Inc.)

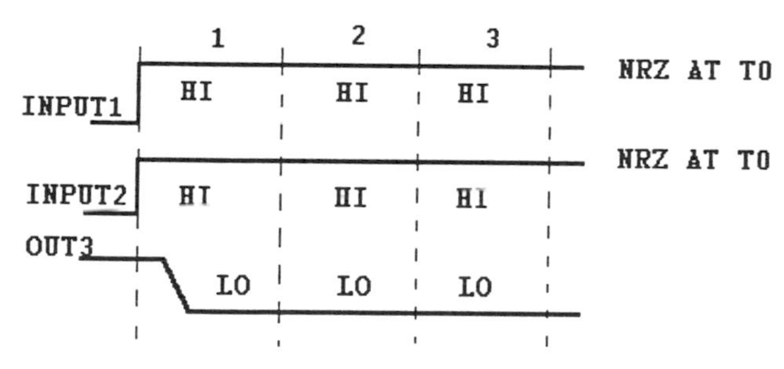

Figure 2.18 Two-input NAND gate timing diagram using NRZ format at t_0.

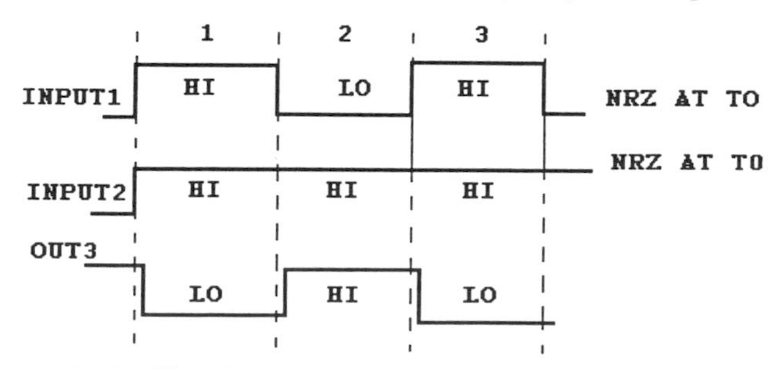

Figure 2.19 Two-input NAND gate timing diagram using NRZ format.

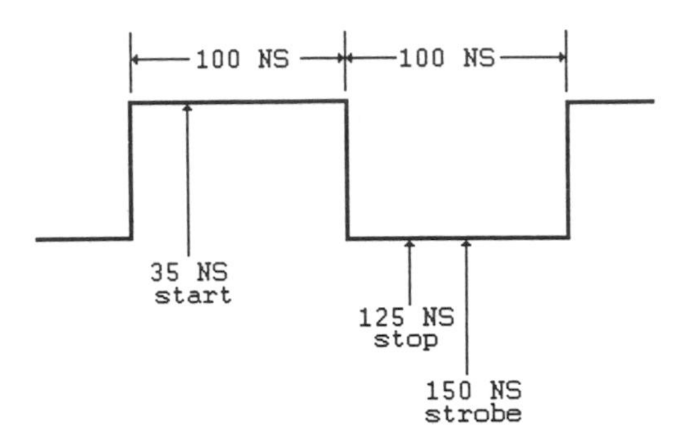

Figure 2.20 Example of measurement conditions.

Table 2.1 Sample parameters arrangement for functional test of a quad NAND gate device

DUT Pin #	1	2	3	4	5	6	7	8	9	10	11	12	13	14
ATE Pin #	7	8	9	10	11	12	13	14	15	16	17	18	19	20
Pin Def.	A1	B1	Y1	A2	B2	Y2	G	Y3	A3	B3	Y4	A4	B4	V_{CC}
1 Device	\	H	M	H	H	X	G	X	H	H	X	H	H	V
2 Truth	H	\	M	H	H	X	G	X	H	H	X	H	H	V
3 T	H	H	X	\	H	M	G	X	H	H	X	H	H	V
4 A	H	H	X	H	\	M	G	X	H	H	X	H	H	V
5 B	H	H	X	H	H	X	G	M	\	H	X	H	H	V
6 L	H	H	X	H	H	X	G	M	H	\	X	H	H	V
7 E	H	H	X	H	H	X	G	X	H	H	M	\	H	V
8	H	H	X	H	H	X	G	X	H	H	M	H	\	V

M = AC measurement; X = don't care; \ = negative-going pulse; G = ground; H = high.

3. The AC comparator voltage level (upper and lower trip point) for propagation delay and rise and fall time must be known. (The polarity of measurement reference points.)
4. The value of load in terms of current as I_{OH} and I_{OL} must also be specified.
5. Timing, including total cycle time, start time, and stop time must be specified. Time between start and stop must be smaller than the total cycle time.

The next step is to provide a functional sequence for AC testing. The repetition of these functional sequences, if selected correctly, should repeat the desired output states for time measurement. A typical functional truth table for a quad two-input NAND gate chip is shown in Table 2.1. The table reveals the following information.

1. While inputs A1 and B1 in operand lines 1 and 2 are stimulated alternately, the gate output pulse is measured (refer to the device specification on the manufacturer's data sheet in Fig. 2.14).

2. The I/O of the first gate is then unstimulated when the functional AC measurement for the next gate is performed on lines 3 and 4.

Since the operation of ATE hardware is controlled by the software commands, the relation between both must be maintained. For example, the test system must know that input 1 and input 2 of the DUT are connected to pin electronic cards number 7 abd 8 of the ATE and that the high and low input voltages are applied to the input pins as described in the next section.

2.7.4 Generic Algorithm for AC Test Software

An example of the method of using software commands to facilitate propagation delay and rise/fall time tests is given here. These commands are not given in any specific computer language but can be used as a guide for any particular test command within any ATE commands words. With respect to Figure 2.20 and 2.21 and Table 2.1, the following generic software commands may make the procedure more understandable.

1. Pin map or channel definition. In this section of the test software every pin of the DUT is assigned to a pin electronic of the test system located on the ATE test head. The test head is the main part of any ATE accessible to the test engineers where pin electronic cards are located.

A *loadboard* is a piece of test hardware that connects the device under test with the ATE's internal capability. A loadboard may have a socket to hold a fabricated device or it may be prepared with appropriate wiring and connectors to facilitate wafer sorting through a probe card or final test through an automatic handler. *Pogo pins* are spring-loaded contacts that are the I/O of the pin electronic cards (PECs) that are normally located inside the test head of ATEs and accessible to the user. The PECs contain the forcing and measuring circuits of the ATE. The number of PECs can vary up to the maximum of the ATE's pin count handling

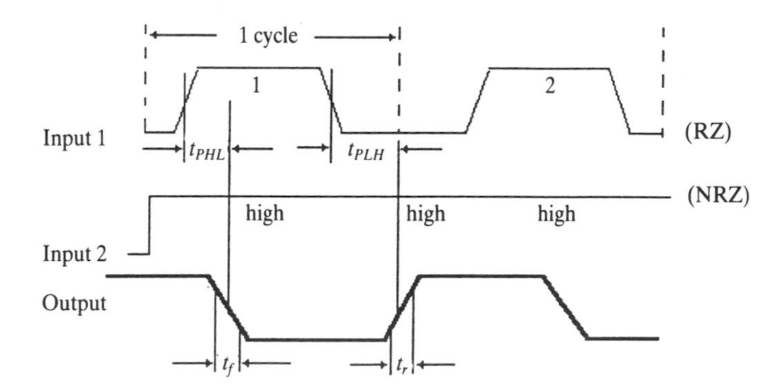

Figure 2.21 Input/output waveforms for a two-input NAND gate using RZ and NRZ wave formats.

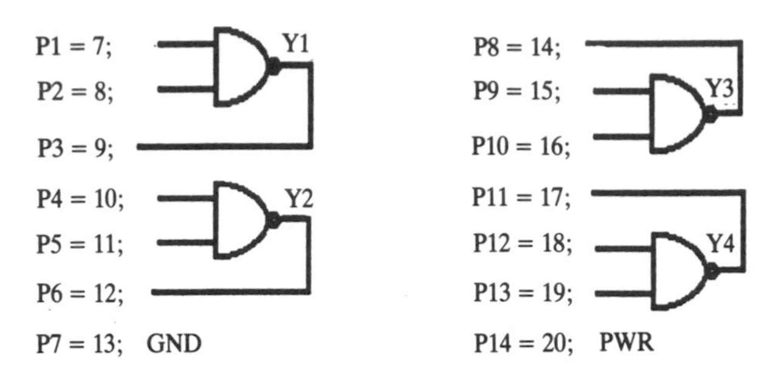

Figure 2.22

capability. Pogo pins connect the DUT pins to the force and measurement units of the ATE through the loadboard. The related NAND gate's I/O is also illustrated in Figure 2.22 to clarify discussion of the test procedure. The P_n represent the pin electronics located on the test head of the automatic test equipment. These pins are under the control of the test engineer and should be assigned and actually connected to the device I/O pin. Once pin assignment is made it remains the same in the program until changed again. These pins can be called by any name, but whatever name is used in the program should not conflict with the reserved words of the operating system of the test equipment's computer.

2. Since pins no. 7 and no. 14 of the device are connected to ground and the supply source of the system, the related required conditions for ground and V_{CC} must be maintained as follows:

"force 0.0 volts clamp 250 mA to pin 7" (1)

"force 5.0 volts clamp 100 mA to pin 14" (2)

In the first statement, 0 V is applied to pin no. 7 (GND pin) of the device allowing it to sink not more than 250 mA current to ground. In the next statement, 5 V (biasing voltage) is applied to pin no. 14 (V_{CC} pin) of the DUT with a current of not more than 100 mA entering to the device's power pin. The values of different voltages or currents can also be assigned by variable names. All test limits and parameters are best declared in the variable declaration section of the test program. This is an example of efficient software writing in which changes can be made on any variable only on one line of the software without needing to search throughout the program.

It is important to select the ground pin clamp current carefully. The allowed current sinking capability of the ground pin must be at least equal to the sum of all the allowed currents entering the device, through both the power and input pins.

3. Set the input and output reference voltages.

"set reference voltage hi 3.0 volts, lo 0.0 volts;" (3)

This statement sets the high and low reference voltage to the DUT inputs. ATE drivers are now set for the input comparison.

"set ref1, ref2 for prop delay 1.5 volts;" (4)

This statement specifies the low trip point of the input pulse for propagation delay measurement. The ATE comparator is set for low trip point voltage. Ref1 and ref2 use any reserved word that specifies the reference points on the input and output pulse, respectively.

"set ref3 for rise time hi 2.8 volts;" (5)

This statement assigns 2.8 V for the upper edge of the output pulse for rise time measurement. The ATE comparator is set for the upper edge of the output voltage. Ref3 is any reserved word that specifies the upper measuring point (usually 90 percent of the total pulse amplitude) of rise time.

"set ref4 for fall time 0.4 volts;" (6)

This statement assigns 0.4 V for the lower trip point of the fall time measurement. A comparator is also set for comparing the low trip point of the output voltage for pass/fail purposes only. Ref4 is any reserved word that specifies the lower measuring point (usually 10 percent of the total pulse amplitude) of rise time.

4. Connecting the programmable load to the gate, outputs.

"connect I_{OL} of 2.0 mA at pins 3, 6, 8, 11;" (7)

This statement connects a load equivalent to drawing 2.0 mA out of the gate's outputs. Refer to the programmable loads section at the end of this chapter.

5. Setting cycle time and functional patterns. This command controls the operation of the timing generator to generate the desired waveform.

"set TG cycle 200.0 ns, start at 35.0 ns, stop at 125.0 ns, strobe at 150.0 ns;" (8)

Usually the generation of the cycle time and comparing the measured timing parameters is performed by a different timing module. This information is derived from Figure 2.20.

"set format R0 at pins 1, 2, 4, 5, 9, 10, 12, 13;" (9)

ATEs usually have a variety of wave formats for forcing the input wave and comparing the measured output parameters. The RZ, NRZ and R0 are widely used in most applications.

The R0 format with the specified cycle time from the previous command line is applied to the gate's inputs. In fact, this statement links the general-purpose timing generator to the test channels through the device pin.

"set ref1, ref2 at pins 1, 2, 4, 5, 9, 10, 12, 13;" (1)

The amplitude of the voltage assigned to the timing cycle is applied to the gate's inputs.

"connect patterns to pins 1, 2, 3, 4, 5, 6, 8, 9, 10, 11, 12, 13;" (11)

This command line connects the device truth table contained in Table 2.1 (lines 1–8) to all the inputs/outputs of all the gates. Notice that the gate output functional response is controlled by the Hi and Lo states given in the device truth table. The cycle time obeys the rules set in that table.

6. The measurement and the required commands will be stated thereafter. Note also that every line in the truth table is dedicated to a unique gate response familiar to the ATE. For example, the time between high-going input pulse to low-going output pulse between pin no. 1 and pin no. 3 of the device is stated using the first line of the truth table (Table 2.1). This propagation delay can be measured within a specified timing range set by the test engineer.

7. The following hints may clarify the procedure for measuring rise and fall time. Rise and fall time of a pulse is measured from 10 percent to 90 percent of the output pulse. The two percentage figures must be applied to the low and high edges of the output pulse. If the lower edge of the output pulse is 0.2 V and the upper edge is 4.8 V, the calculation to determine 10 to 90 percent of the rising edge will be as follows:

$$A = \frac{4.8\,\text{V} \times 10}{100} + 0.2\,\text{V}$$

$$B = \frac{4.8\,\text{V} \times 90}{100} + 0.2\,\text{V}$$

A is the lower edge and B is the upper edge where the rising time must be measured. There are specific reserved words for A and B in our example in any ATE system.

A typical statement showing this time range is as follows:

"measure time pin 9 and pin 12 range 150.0 ns using line 1 to 2" (12)

This statement shows the time range in which the measurement is going to be made as well as the number of specific functional patterns (lines 1 and 2 of Table 2.1). The measurement is made at two outputs only. The measured value can then be compared with the predefined high and low limits for that particular parameter; pass and fail can then be determined through the software codes.

The value of load connected to the output must be declared as variable and should be commanded. The combination of the above arranged function truth table and the test software must be translated to machine codes acceptable to the test system (this is done by the system through compilation).

8. A similar procedure is also applied for propagation delay measurement. The rise/fall test procedure needs a minor modification since the propagation delay is measured from 50 percent of the rising or falling edge of the input pulse to 50 percent of the rising or falling output pulse. A and B are now measured from 50 percent of input and output pulse. A generic software code for propagation delay measurement is as follows:

"measure time pin1 to pin3 range 150.0 ns using functional lines 1 and 2" (13)

It is clear that the measurement is made from 50 percent of the rising or falling edge of the input pulse at pin no. 1 to 50 percent of the rising or falling edge of the output pulse at pin no. 3 of the Y1 gate in our example, refer to code (4). The pass and fail of the measured limit will be stated thereafter.

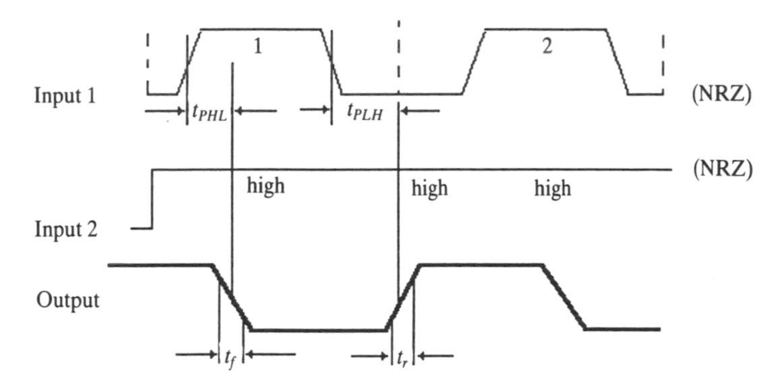

Figure 2.23 Two-input NAND gate timing diagram using only NRZ wave format.

The organization and complexity of any software program are solely dependent on the test program developer. In any ATE there are specified reserved words burned into the system *read only memory* that enable test engineers to manipulate internal system hardware. Words such as "force," "measure," "connect," "disconnect," "amps," "volts," "clamp," etc., that are common in any ATE system are nonexistent in ordinary high-level or low-level computer languages.

The RZ format is applied at input 1 and NRZ is applied to input 2. The output transition occurs at every clock cycle. The range of time to measure rise and fall or propagation delay time on any average ATE is within milliseconds.

2.7.5 Algorithm for Functional Test

Lines 1–8 of Table 2.1 represent the expected correct functionality of the device in our example. This information is used for functional testing as well. A functional test is the verification of the truth table by the operation of the device. The number of functional patterns that a device should be tested for is decided by the test engineer and is dictated by two factors: First, the greater the number of functional patterns, the longer the test time; second, consideration must be given to the probability of device operation failure for any particular sequence that may cause the worst case. The latter is not an easy task to deal with, especially when testing complex devices.

Most of the steps that have been explained for propagation delay or rise/fall testing will be used here with the following exceptions:

1. The high and low level of the output voltage must be given to the system comparator.
2. The gates' outputs are not usually loaded. The output voltage will be compared with the reference levels instead of being measured.

In ALL ATE systems there is a specific reserved word for functional test performance. One example is

"functional test at 1 to 8"

This executes lines 1 through 8 of Table 2.1 for device functional verification.

The pass and fail decision, beginning with "if" and "then" statements, decides whether the device has performed the predicted functions correctly.

2.7.6 DC or Static Parametric Tests

All tests that fall into this category involve voltage or current measurement, either forcing current and measuring voltage or vice versa. The DC test procedure outlined in this book can be done by individual electronic tools in a laboratory, or by an ATE through an organized test program. In essence, DC testing is the same as the functional test with the exception that the input/outputs of gates are connected to the DC measurement system instead of being attached to the comparators. For this reason, the input/outputs of all the respective gates must be connected to the DC measurements units through the test software. These DC measurement units are capable of forcing voltage to a pin and measuring the current passing through, or vice versa, and test the measured value against the limit. For the DC parametric test, the following basics must be defined in the test program:

1. Power and GND pin of the device must have proper voltages applied.
2. If other biasing voltage is required, it must be applied to the defined pin.
3. The value of current that is to be forced into or pulled out of the device output must be defined.
4. The required voltage or necessary current to be applied to the device terminal must be defined.
5. The limits that the measured parameter are compared with must be defined.
6. The pass and fail determination based on the result in Step 5 to determine the bin number.

2.7.7 Programmable Loads

The following parameters for the specific DUT must be known in order to use the various programmable loads available in any ATE:

- The I_{OH} for the sourcing current
- The I_{OL} for the sinking current
- The threshold voltage at which I_{OH} switches to I_{OL}

These program-controlled dynamic loads help test engineers to measure the output voltage at a variety of loads, as well as the fan-out drive capability and the output transitional response of DUTs.

The characteristic of voltage vs. current of the programmable load in Figure 2.25 is shown in Figure 2.24. The transitional voltage is about 1.5 V.

A typical termination circuit or loading network is shown in Figure 2.25. Schottky diodes are used in these types of networks. Schottky diodes are metal

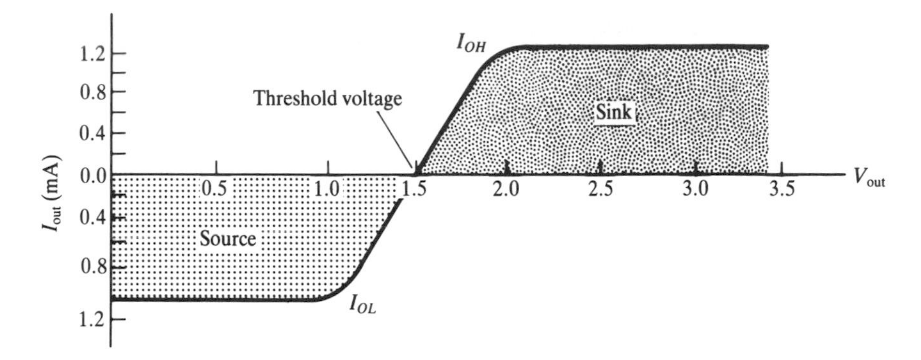

Figure 2.24 Voltage vs. current of programmable load.

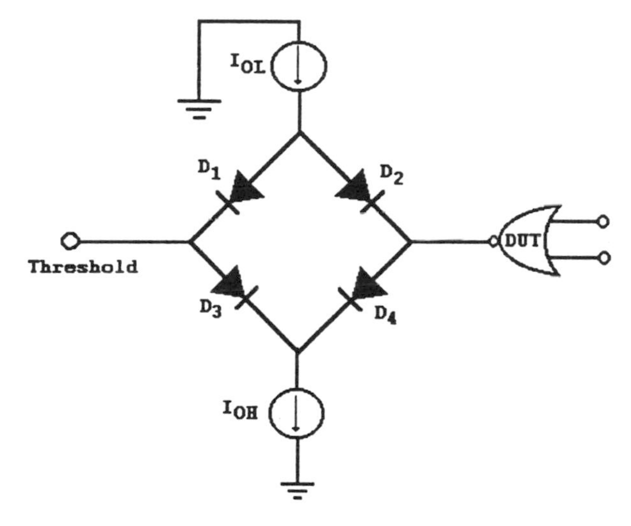

Figure 2.25 Schematic of a termination circuit.

semiconductor diodes, generally incorporated into ICs as clamps between the base and emitter of transistors, to prevent saturation and for fastest recovery. For the same forward current, the voltage drop across such a diode is less than that of a conventional semiconductor diode. In a Schottky transistor, diode and transistor may be produced at the same time.

The termination circuit operates as follows.

1. If the DUT is in high-impedance mode or if the socket is empty (no DUT), all of the diodes conduct and the voltage to DUT will be about 1.4 V, which is the sum of the two diodes' threshold voltages (0.7 V + 0.7 V). This is V_{OL}.
2. When the device output voltage is below threshold voltage, the DUT sinks I_{OL} through diode D2 because D1 and D4 are off.

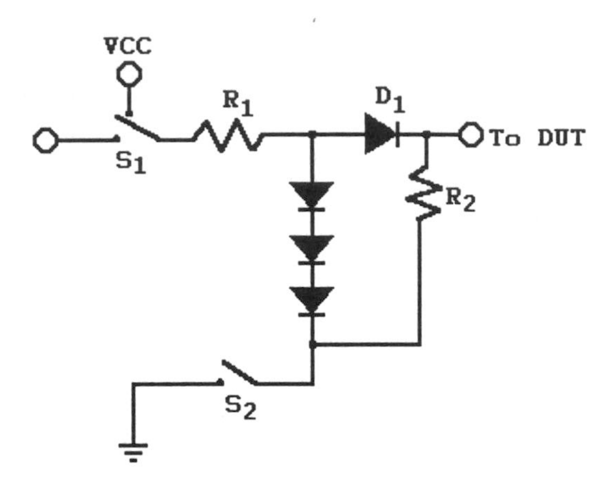

Figure 2.26 A typical load circuit.

3. When the device output is above threshold voltage, the device sources I_{OH} through D4 because D2 and D3 are off.

The manufacturer's device data sheet specifies the values of I_{OH}, I_{OL}, and the threshold voltage, which is usually halfway between V_{OL} and V_{OH} levels. A typical load circuit is shown in Figure 2.26. Switches S1 and S2 are used as pull-up and pull-down loads in high- or low-impedance mode of the DUT. The following situation exists when switches S1 and S2 are both closed:

1. All diodes conduct when the socket is empty. The terminal will have the voltage equivalent to a drop across all diodes (1.4 V).
2. When the DUT voltage is below V_{OL}, only D1 conducts and I_{OL} sinks through resistor R_1 by V_{CC}.

$$I_{OL}=\frac{(V_{CC}-V_{OL}-0.7\,\text{V})}{R_1}$$

3. When the DUT voltage exceeds 1.4 V or is at V_{OH}, the current is sourced through resistor R_2 and

$$I_{OH}=\frac{V_{OH}}{R_2}$$

4. Switch S2 is opened for the TPZL (low state propagation delay at three-state mode) test and the load circuit of Figure 2.26 reduces to the circuit shown in Figure 2.27.
5. Switch S1 is open and S2 is closed for the TPZH (high state propagation delay at three-state mode) test. The loading network consists of just the resistor R_1 in this test.

Figure 2.28 shows a fast, high-performance and low-cost tester for digital LSI and VLSI as well as programmable logic devices. It runs on a VENIX operating

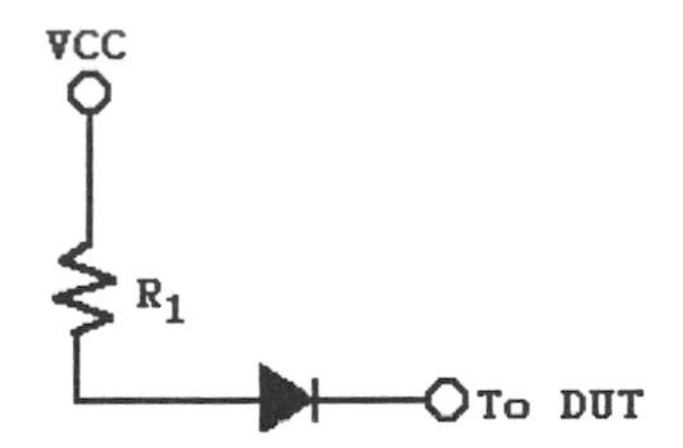

Figure 2.27 Three-state propagation delay test load.

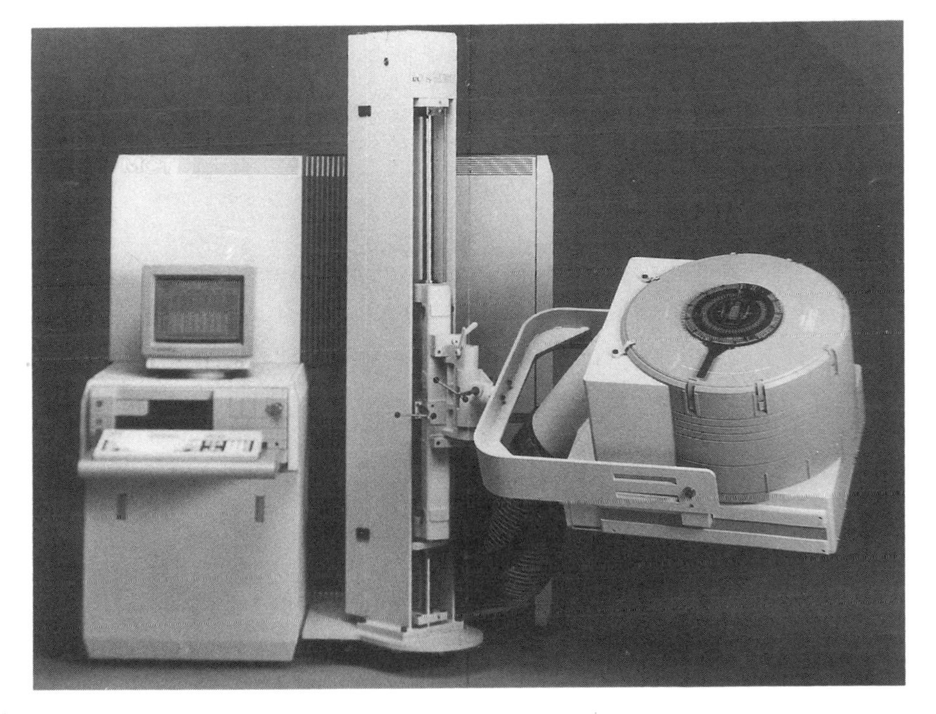

Figure 2.28 Fast, accurate, and affordable automated test equipment for testing SSI, MSI, LSI, and VLSI semiconductor digital circuitry. (Courtesy of Micro Component Technology, Inc.)

system, and has an overall testing accuracy of ± 500 ps with ± 150 ps accuracy in AC parametric measurement. Each pin on the test head, or pin card (which can hold up to 128 pins), operates with full function of precision measurement units (PMU) to perform parallel DC measurement. This tester is capable of handling UNIX shell programming tools as well as networking through interface hardware. It runs up to 100 MHz per pin drive data for high-speed device functional testing with 200 MHz clock data per pin. It is capable of generating wave cycles as low as 20 ns period and 20 ps timing resolution. Every pin card has up to 256K vector memory with ±10 V dual power supply and ±500 mA force/measurement capability. The dual power supply can source/sink at values up to 2 amps. This ATE can perform two different methods of AC measurement (time measurement)—one-shot and averaging. The system is also capable of in-phase (synchronized) or out-of-phase (asynchronized) pulse measurement.

References

1. *Semiconductor Test Solution* (*STS*), Part No. 93-0048, Rev A.
2. *MCT Series 2000 Application Notes Manual.*
3. *Sentry Series 80 Mixed Signal ATE Application Notes.*
4. Digital Systems Division, *Teradyne J283 System Manual.*
5. Graf, Rudolf F., *Electronic Data Book*, 4th ed., TAB Professional and Reference Books, 1985.

CHAPTER 3

Digital Logic Test

Introduction

Digital circuits, unlike analogs, deal with discrete pulses known as digits. In these types of networks, discrete signals in the form of either pulse or voltage levels are manipulated as discrete quantities. These discrete quantities are the representation of logic high or low. The logic high level is known as binary "1" and the low level is called binary "0", or On and Off, respectively. Relays and switches are the elements that are primarily implemented in logic networks. These switches are known as logic gates. Yet there is no known state between the high (binary "1") and low (binary "0") levels in digital circuit operation. As digital test engineers, we are dealing with either logic high or logic low at the input and expecting similar responses at the output of such networks.

This chapter contains the fundamentals of semiconductor digital logic network testing. The purpose and definition of each test are given before the description of the procedure. The direction of current measured is shown in an I/O diagram to make it clear for the test engineer, product engineer, or technician.

Topics

- Input/output structure of TTL or CMOS device
- Step-by-step procedures for all DC, AC, and functional tests that exist in any semiconductor digital logic device

  3.1. Continuity test
  3.2. Shorts test
  3.3. Functional test
  3.4. I_{CC} test
  3.5. Breakdown voltage (VB) test
  3.6. V_{OH} test (high level output voltage)
  3.7. V_{OL} test (low level output voltage)
  3.8. I_{IL} test (low level input current)
  3.9. I_{IH} test (high level input current)

3.10. IOZH test (high level three-state output current)
3.11. IOZL test (low level three-state output current)
3.12. I_{OL} (saturation) test (low output saturation current)
3.13. I_{OH} (saturation) test (high output saturation current)
3.14. Voltage hysteresis (V_{HYS}) test
3.15. A binary search algorithm
3.16. AC testing description
3.17. Propagation delay test (t_{PLH}, t_{PHL}, t_{PHZ}, t_{PLZ})
3.18. Output pulse width (t_W) test
3.19. Rise and fall time test (t_R, t_F)
3.20. Setup time test (t_{SU})
3.21. Hold time test (t_H)
3.22. Removal time (t_{REM}) or recovery time (t_{REC}) test
3.23. Noise in digital circuits—ground noise in digital circuits; reflection (ringing), crosstalk

- Description of all noise that is common in digital circuits, including practical ways to identify the noise, a nonmathematical method for suppressing it, and related schematics and figures.

The procedures that are outlined in this chapter, along with the input and output relationship between the forcing and measuring parameters, can be performed in a laboratory having the proper equipment or using an ATE through an organized test program. Each test described in this chapter starts with the definition and intention. Some tests, due to their complexity or importance, have been described in more detail to help the reader thoroughly understand the tasks involved. Tests such as voltage hysteresis and pulse width require in-depth detail of the nature of the parameters as well as the in-depth test procedure.

The input/output structure of a TTL and of a CMOS device is given in Figure 3.1. The internal circuitry between input and output can be very complex. The diagram represents three inputs and one output for the bipolar and one I/O for high-speed complementary (MOS) (HC, HCT) types with protection diode structures at the input and the output. These diode structures protect the internal device circuitry from damage by static charges during product handling. The main device circuitry is located in the box named the logic circuit, which can be any type of complex structure. The device internal circuitry is not a matter for our concern, as long as the functionality of the device can be demonstrated by the application of the related truth table to the input(s) and observing the predicted response at the output.

There are only three distinguishable groups of tests applied on semiconductor devices:

- Functional tests
- Static or DC parametric measurement (current and voltage)
- AC or dynamic testing, which is basically the measurement of time and time-involved operations.

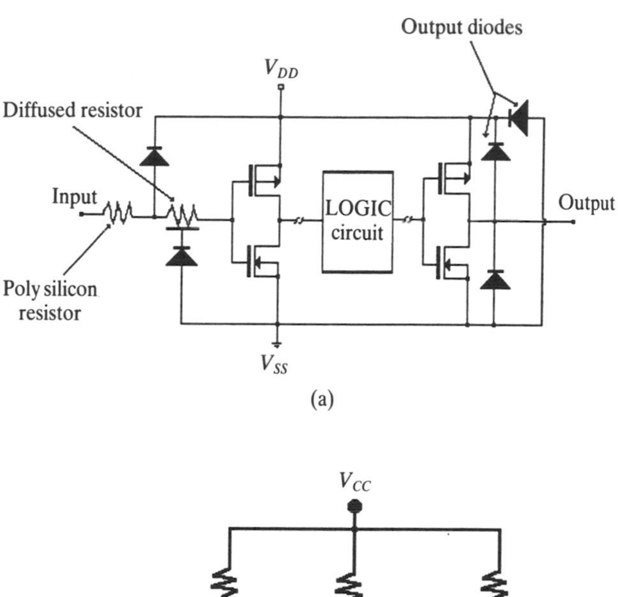

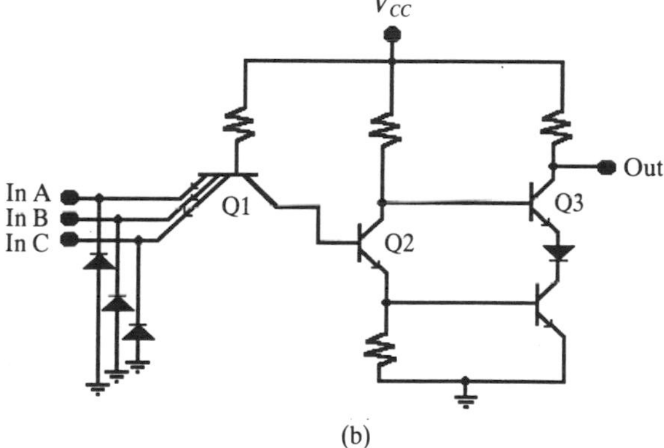

Figure 3.1 Input/output structure of (a) TTL and (b) MOS devices.

The combination of all three groups may guarantee the operation of a device according to the design intention stated in the related specification sheet.

The sequence of tests in a test program is important. There are some tests that must be performed whether they were suggested or not; they give an indication that other tests may proceed. For instance, the connectivity test gives an indication of whether there is a connection between the DUT and the test system. The short test proves that the device pins are not internally or externally shorted to one another. These tests verify the validity of DUT and test hardware for further and more complex tests.

All tests that are required for any logic device are given in this book, regardless of device type or manufacturer. The best effort has been made to make the matter as clear as possible so that it can be understood by inexperienced test engineers. Test limits are not given in any of the following test procedures. There are, however, a few instances where the general values of V_{CC} or current limits are given, for comparison purposes only.

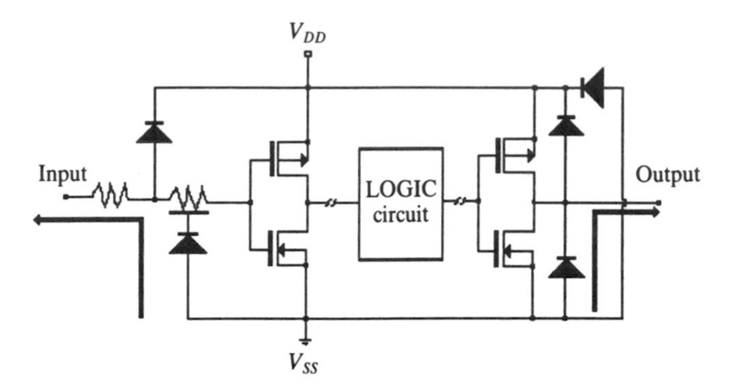

Figure 3.2 Lower continuity test.

3.1 Continuity Test

Purpose

This test verifies that the device pins are connected to the test system through pogo pins.

Procedure

A current (its magnitude is dependent on the device types and process) is applied to all the I/O pins and the resulting voltage is measured at every pin. The direction of current is shown in Figure 3.2. The ground (GND) or power pin should be maintained at 0 V. This test also ensures the existence of the protection diodes at every I/O pin as well as the device input resistance.

1. Apply 0 V to V_{DD} or V_{CC} and GND pins.
2. Force a defined current to each I/O pin.
3. Measure voltage at the I/O pins and V_{DD}/V_{CC}.

3.2 Shorts Test

Purpose

This test checks whether any pin of the device is shorted to another, either externally or internally.

Procedure

1. Apply a nominal voltage to GND, V_{DD}/V_{CC} pins.
2. Apply 0 V to the desired pin while all other I/O pins are at a defined voltage. The magnitude of the voltage varies in different IC technologies.
3. Measure voltage level at all I/O pins except the pin being tested in the second part of Step 2. The measured voltage should show whether that particular pin is shorted with others.

4. Repeat the same procedure until all I/O pins in turn go through 0 V application.

3.3 Functional Test

Purpose

Functional testing is the application of a train of pulses to the input pins of the DUT and checking the outputs for their response. The input stimuli and the output sensing are determined by the logic function (based on the suggested truth table or timing diagram) of the device given by the manufacturer.

Functional tests can be grouped in many ways depending on V_{DD}/V_{CC} level or the gross and threshold V_{IH} and V_{IL}. A functional test performed using threshold V_{IH} and V_{IL} is used to prove the sensitivity of the device to minor input level variation (noise immunity test).

MOS devices are tested under no load. TTL devices are loaded and maximum and minimum V_{CC} is used

Procedure

1. Force V_{DD}/V_{CC} to the power pin and 0.0 V to the GND pin.
2. Perform functional test using the prescribed truth table and gross V_{IH} and V_{IL} as high and low input voltages respectively. The functional pattern(s) can be generated automatically or manually created in almost any ATE. Input and output must be specified.
3. Monitor the state of the outputs (V_{OH}, V_{OL}) according to the given specifications. A functional test is a go/no go or pass/fail test. The output level voltage is compared, not measured.

3.4 I_{CC} Test

Purpose

This test measures current passing through the V_{DD}/V_{CC} pin when the device is in different output conditions. These conditions will have different current readings. The output conditions can be one of the following cases, depending on the device function:

- High state
- Low state
- Three-state (high-impedance) condition

No device is loaded while this test is in progress.

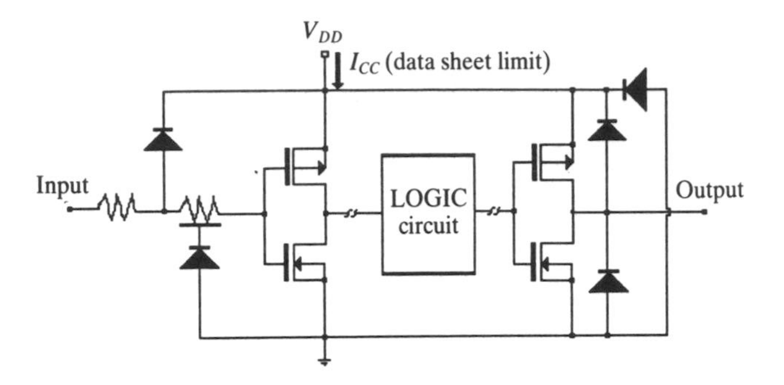

Figure 3.3 I_{CC} test.

Procedure

1. Apply maximum V_{CC}/V_{DD} to the power and 0 V to the GND pin.
2. Apply gross (nonthreshold) V_{IH}/V_{IL} to the functional patterns (truth table).
3. Measure I_{CC}/I_{DD} at the power pin for all possible patterns.

The test program must contain all the possible patterns that prove the total logic function of the device. The I_{CC}/I_{DD} limit is determined and stated in the manufacturer's data sheet for any particular device. Figure 3.3 shows the direction of current.

3.5 Breakdown Voltage (VB) Test

Purpose

This test checks whether the V_{DD}/V_{CC} pin will break down when maximum supply voltage is applied to the power pin. This maximum supply voltage (also known as the Zener voltage) is given in the device data sheet and it is neither applicable for the ordinary DC testing nor should it be used to operate the device.

TTL devices can operate as high as 7.0 V and high-speed MOSFETs can handle up to 12.0 V.

No load should be used in this test.

Procedure

1. Force supply voltage to V_{DD}/V_{CC} and 0 V to the GND pin.
2. Perform the same test as I_{DD}/I_{CC} for all possible output conditions.
3. Measure current at V_{DD}/V_{CC} pin.

The measured current must be within the manufacturer's suggested limits.

3.6 V_{OH} Test (High Level Output Voltage)

Definition

V_{OH} is the minimum high level voltage to which the device output will fall for a given current load when V_{DD}/V_{CC} is minimum.

Purpose

This test guarantees the output load drive capability fan-out of the device when the output is high.

TTL and MOS devices are loaded for this test. The loading values are defined by the manufacturer.

Procedure

1. Apply V_{DD}/V_{CC} to power pin and 0 V to the GND pin.
2. Using the information in the device truth table or timing diagram, connect the functional sequences that change the state of the output to high logic level to the input pin(s).
3. Measure the output voltage when the specified load is connected.

The measured value must agree with the manufacturer's suggested limits.

3.7 V_{OL} Test (Low Level Output Voltage)

Definition

V_{OL} is the maximum low level voltage to which the device output will rise for a given load when V_{DD}/V_{CC} is minimum.

Purpose

This test guarantees the capability of the device to handle load fan-out when the output is at logic low.

All TTL and MOS devices are tested under a specified load.

Procedure

1. Apply V_{DD}/V_{CC} to the power pin and 0 V to the GND pin.
2. Apply appropriate functional sequence (pattern of truth table) to the inputs that sets the output to logic low state.
3. Measure resulting voltage at the output pin when an applicable load is connected.

The measured voltage must be within the manufacturer's suggested limits.

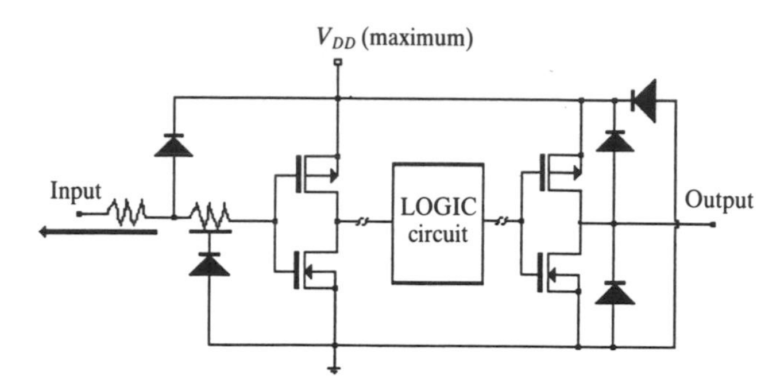

Figure 3.4 I_{IL} test (low level input current).

3.8 I_{IL} Test (Low Level Input Current)

MOS devices, unlike their TTL counterparts, maintain very low input current leakage, on the order of picoamps. This is due to reverse leakage (minority charges crossing the junction) of the input protection diodes.

Purpose

This test checks the current passing through the input protection diodes for all the inputs that set the device output to logic low level. The existence and the value of the base resistor can also be determined by this test. The direction of current is shown in Figure 3.4.

No load is required for either TTL or MOS devices.

Procedure

1. Force V_{DD}/V_{CC} to the power pin and 0 V to the GND pin.
2. Apply the maximum specified V_{OL} on all input(s).
3. Measure the resulting current at the input(s).

The measured voltage must agree with the suggested limits.

3.9 I_{IH} Test (High Level Input Current)

Purpose

This test guarantees the β value of the input transistor and its reverse emitter-to-collector operation. This is the total current passing through the input protection diodes for all the input conditions that set the device output in logic high level. The direction of current is shown in Figure 3.5.

No load is required for either TTL or MOS devices.

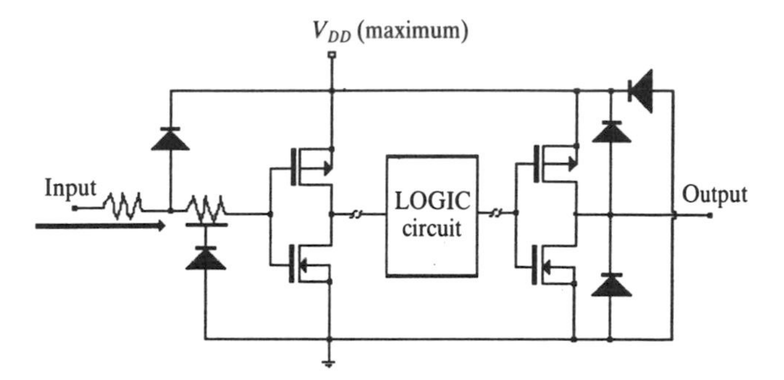

Figure 3.5 I_{IH} test (high level input current).

Procedure

1. Force V_{DD}/V_{CC} to the power pin and 0 V to the GND pin.
2. Force the specified V_{IH} on all the input pins.
3. Measure current at input(s).

The measured current must stay within the suggested limits.

All devices having three-state condition undergo the same procedure for I_{IL} or I_{IH} tests. However, this condition must be provided prior to the test. The state that is maintained by a particular pin on such devices is dependent on their design. In this state the device output is in high-impedance condition and theoretically should not draw any current. Three-state gives priority to other devices sending information through a common bus.

3.10 IOZH Test (High Level Three-state Output Current)

Purpose

This test measures the output current of the device at high output level, when the device is in high-impedance state (three-state). A specified pin on any device having this state must be stimulated. A typical buffer has a high input and a low output impedance when used for isolation of two networks. The direction of current is shown in Figure 3.6.

Neither TTL nor MOS devices should be loaded in this test.

Procedure

1. Force V_{DD}/V_{CC} to the power pin and 0 V to the GND pin.
2. Force the device output(s) to high-impedance state.
3. Apply proper logic level at the input(s) that sets the output(s) at logic high state.

The measured value must be within the specified limits.

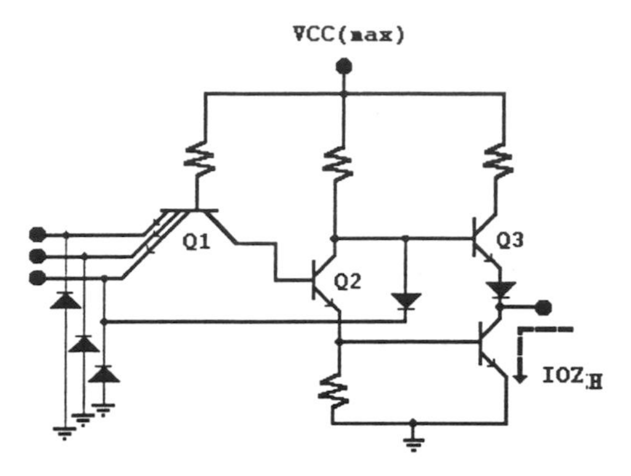

Figure 3.6 IOZH test (high level three-state output current).

3.11 IOZL Test (Low Level Three-state Output Current)

Purpose

This test measures the output current of the device at high-impedance (three-state) condition when the output is at logic low level. A proper stimulus to the three-state pin creates this condition. The direction of IOZL current is shown in Figure 3.7.

Neither TTL nor MOS devices should be loaded for this test.

Procedure

1. Force V_{DD}/V_{CC} to the power pin and 0 V to the GND pin.
2. Force the device output(s) to the high-impedance state.

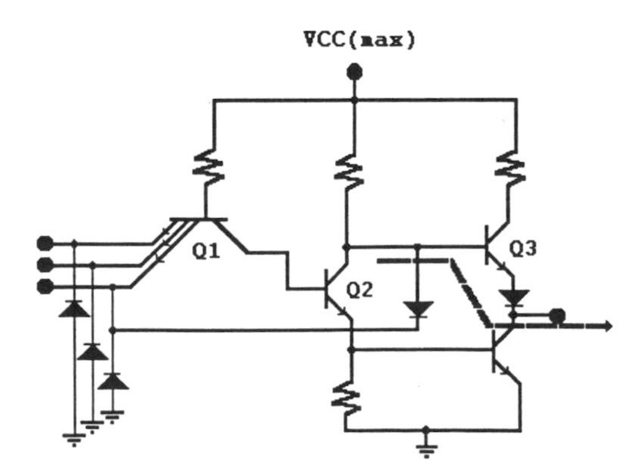

Figure 3.7 3.7 IOZL test (low level three-state output current).

3. Apply the proper functional pattern (truth table) to the input(s) that sets the output(s) at logic low state.
4. Measure current at the output pin(s).

The measured voltage must be within the specified limits.

3.12 I_{OL} (Saturation) Test (Low Output Saturation Current)

Purpose

I_{OL} or I_{OL} (saturation) are similar parameters by which the absolute maximum current sinking capability of the device is measured. This test is done by placing the device output(s) in logic low level while measuring the resultant current flow into the device output. The direction of current is shown in Figure 3.8.

Neither TTL nor MOS devices should be loaded in this test.

Procedure

1. Force V_{DD}/V_{CC} to the power pin and 0 V to the GND pin.
2. Apply proper voltage (high and low logic sequences from the device truth table) at the input(s) to set the device output into low state.
3. Measure current at the output pin(s).

The measured current must agree with the specified limits.

3.13 I_{OH} (Saturation) Test (High Output Saturation Current)

Purpose

I_{OH} or I_{OH} (saturation) are similar parameters by which the absolute maximum current sourcing capability of the device is measured by placing the device output(s) in logic high level. The direction of current is shown in Figure 3.9.

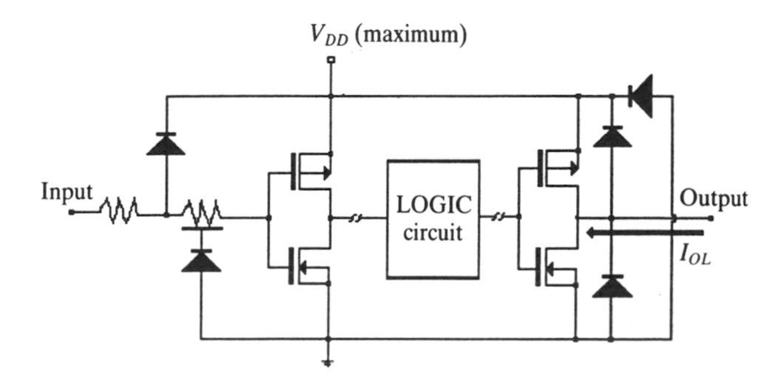

Figure 3.8 I_{OL} (saturation) test.

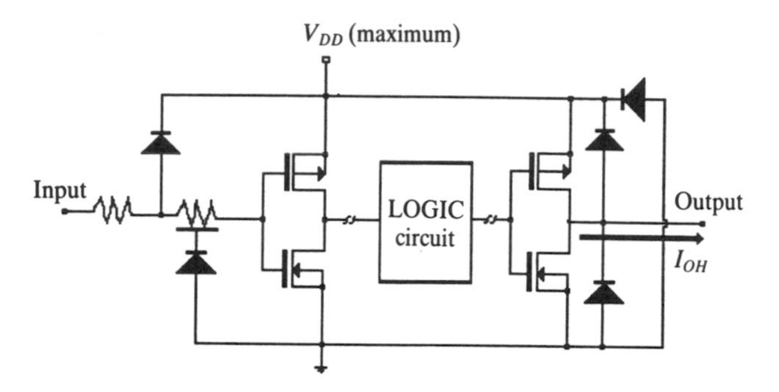

Figure 3.9 I_{OH} (saturation) test.

Procedure

1. Force V_{DD}/V_{CC} to the power pin and 0 V to the GND pin.
2. Apply proper functional patterns at the inputs (related truth table) to set the output in logic high state.
3. Measure current at the output pin(s).

The measured current must be within the specified limits.

Neither TTL nor MOS devices should be loaded in this test.

3.14 Voltage Hysteresis (V_{HYS}) Test

Definition

A hysteresis loop is applied to any pair of variables when their return path is different from the forward. The symbol of this path is shown on Schmitt trigger gates. The lag of the device in responding to defined high and low input voltages is called *hysteresis.* Such devices are used for noise filtering and creating extremely sharp triggering edges from ragged-looking pulses.

These devices trigger to a definite state when the input voltage is above the upper trip point, and remain in that state even though the voltage has dropped below that point. The state of the device will change only when the voltage drops below the lower trip point. Hysteresis in small amounts helps the rapid switching action of internal clocks and covers wide temperature variations. *Voltage hysteresis* is the difference between the positive-going input voltage (upper trip point) when the output switches, and the negative-going input (lower trip point) when the output switches again. This is in fact the difference in value between two threshold voltage tests.

Purpose

This test is performed on Schmitt trigger devices having input voltage hysteresis, and guarantees a defined noise immunity level on the input(s) within the transistor region. The hysteresis test is performed when the device is loaded.

Procedure

Positive-going Threshold Voltage (V_{T+})

1. Force power to the V_{DD}/V_{CC} pin and 0 V to the GND pin.
2. Apply 0 V at the device input(s). Increase the voltage at the desired input pin in very small steps (millivolts) while monitoring the output for response. The input voltage at which the output switches should be recorded, and it must be within the suggested limits.
3. Force 0 V at all the inputs.

Negative-going Threshold Voltage (V_{T-})

1. Force power to the V_{DD}/V_{CC} pin and 0 V to the GND pin.
2. Apply V_{DD}/V_{CC} to all the device inputs.
3. Start decreasing voltage at the desired input while monitoring the output for response. The decrement steps should be in millivolts.
4. Measure the voltage point at which the output has responded. The measured voltage must agree with the suggested limits.
5. Apply V_{DD}/V_{CC} to all the inputs. At this point we have obtained V_{T+} and V_{T-}. Voltage hysteresis is the difference between these two threshold voltages:

$$V_{HYS} = V_{T+} - V_{T-}$$

Figure 3.10 shows hysteresis output vs. input.

The V_{HYS} value must be within the manufacturer's suggested limits. Accurate hysteresis measurement due to step incrementation and decrementation of the

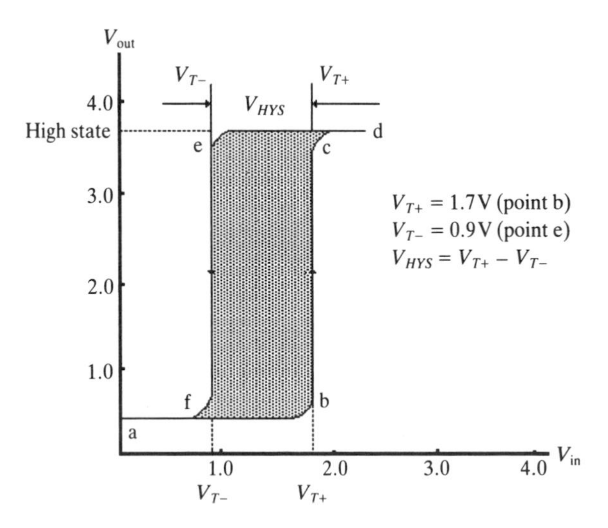

Figure 3.10 Hysteresis output vs. input.

input voltage is time-consuming, which is costly. There are two other methods that can reduce the test time and process drastically: binary search (successive approximation) and sample and hold.

3.14.1 Binary Search (Successive Approximation) Method

The binary search method reduces the test loop to not more than eight, which, in traditional methods, is in the order of tens for a fine measurement. The binary search method is merely a software modification and does not require additional hardware.

A sample of the binary search method is given in Section 3.15. This procedure is written in C and can easily be modified to any ATE software.

3.14.2 Sample and Hold Method

The sample and hold method is the most recent invention in the ATE world. A ramp and hold module is employed to find the input threshold. Ramp generators can be programmed for the ramp rate and threshold. The program controls the ramp slewing and will stop as the input signal matches the programmed threshold voltage.

3.15 A Binary Search Algorithm

```
#include <stdio.h>
#include <ctype.h>
#include <math.h>

/* This function used to manually enter the output voltage for a given input
voltage. The function displays the input voltage and assists the user to manually
enter the output status; i.e., high (h) or low (l). */
int out_put(float of_level, int low_true)
{
     int x,y;
     char c;
     c='x';
     while ((c!='h') && (c!='l'))
{
     if (low_true)
     printf(   "\n\n input voltage is percentf and the previous output is low
     \n",of_level);
else
     printf(   "\n\n input voltage is percentf and the previous output is
     high\n",of_level);
```

```
        printf("enter if output is low(l) or high(h) for the above input volt:");
        scanf("\n percentc",&c);
}
        if (c= ='h')
        return(l);
        else return(0);
}
main()
{
        float max,min,p_high,p_low,delta,f_level,low_hysteresis,high_hysteresis;
        int low_true=1;
        max=5.0; min=0.0, delta=0.1;
        p_high=max;
        p_low=min;
/* p_high and p_low are used as temporary storage for minimum and maximum
voltage; delta determines the accuracy with which the low-hysteresis and
high-hysteresis is found. */
/* find low to high hysteresis*/
        while (p_high=p_low)> =delta
{
        f_level=(p_high+p_low)/2;
        if (out_put(f_level,low_true)= =1)
{
p_high=(p_high+p_low)/2;
/* reset_low(); */
}else
{
        p_low=(p_high+p_low)/2;
}
}
        low_hysteresis=(p_high+p_low)/2;
        printf("     **** the low to high hysteresis percentf volts     ****\n     ",
        low_hysteresis);
/* find high to low hysteresis */
```

```
        low_true=0;
        p_high=max;
        p_low=min;
        while (p_high-p_low)> =delta)
{
        f_level=(p_high+p_low)/2;
        if (out_put(f)level,low_true)= =1
{
        p_high=(p_high+p_low)/2;
/* reset_low(); */
}
else
{
        p_low=(p_high+p_low)/2;
}
}
    high_hysteresis=(p_high+p_low)/2;
    printf("****    the high to low hysteresis percentf volts     ****\n\n    ",
    high_hysteresis);
    printf("****    the differential hysteresis percentf volts     ****\n\n    ",
    low_hysteresis-high_hysteresis)
} /*end of main*/
```

3.16 AC Testing Description

Finite time is required for a transistor to switch. This time affects the speed and the shape of the pulse before and after the switching occurs. *Propagation delay* is defined as the time required from application of an input signal to the resulting change on a specified output. It is usually measured from 50 percent of the amplitude of the applied pulse to 50 percent of the amplitude of the response pulse.

Rise time is normally measured from 10 to 90 percent of the *rising edge* of a pulse. Fall time is the measurement of the time over which a pulse falls from 10 to 90 percent of its amplitude.

The shape of the output pulse is not usually the same as the input signal. The slope of the pulse shows that a longer time is required for the rising edge to reach the "high" logic operating level. This is called *rise time. Fall time* is the time required for the falling edge of the pulse to reach the "low" logic operating level.

The accumulation of propagation delays in a complex circuit represents the

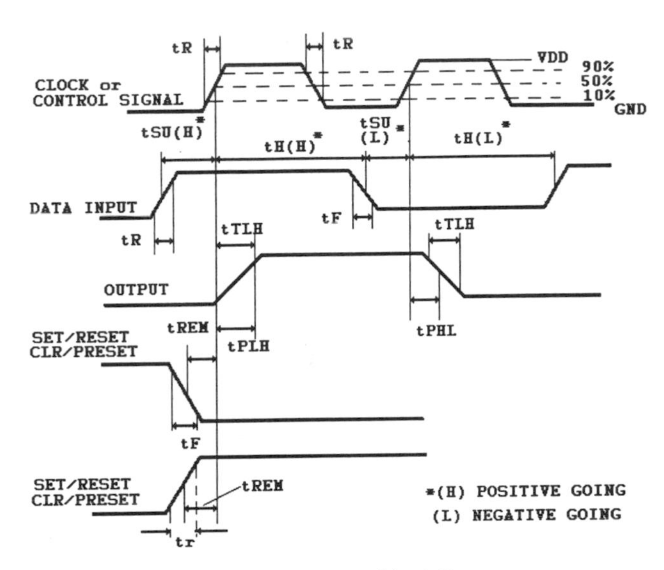

Figure 3.11 Measureable AC parameters.

overall speed of the system. If the propagation delay in a switching transistor is long, the shape of the output pulse will be degraded and the same pulse will appear flat after undergoing a number of switching stages. Propagation delay can be measured from the low-going input pulse to the rising output pulse or vice versa and is the same for three-state devices.

The symbols for propagation delay are t_{PLH} and t_{PHL} where *LH* and *HL* represent the rising or falling edge of the output pulse, respectively. These are shown in the waveform in Figure 3.11.

The symbols for rise and fall time are t_R and t_F, respectively. In the next section the details of and the reasons behind each digital AC measurement are given.

3.17 Propagation Delay Test (t_{PLH}, t_{PHL}, t_{PHZ}, t_{PLZ})

Purpose

This test measures the time between the application of a specified voltage at the input and the resulting change at a specified output. It verifies the switching speed of the DUT. t_{PHZ} and t_{PLZ} are called *disable* time, while t_{PZL} and t_{PZH} are called *enable* time in some semiconductor manufacturers' data books and are referred to as propagation delay in the three-state (high-impedance) condition.

TTL and MOS devices must be loaded in this test.

Procedure

This method for measuring propagation delay can be applied to a device at a laboratory bench when the proper equipment is available, or on an ATE through a test program (refer to Sections 2.7.3 and 2.7.4 in Chapter 2). The nature of the test is the same in either case.

1. Force V_{DD}/V_{CC} to the power pin and 0 V to the GND pin.
2. Apply the proper functional pattern (from the device truth table) to the input(s).
3. Check the state of the output(s) after a proper time delay.
4. The device output must be in the expected logic level.

The functional patterns in Step 2 determine the state of the output with respect to the input level (*LH* or *HL*). In the case of three-state devices, the same procedure applies and the three-state condition must be maintained before the test.

The types and the current sinking capability of the load are specified by the manufacturer.

3.18 Output Pulse Width (t_W) Test

Discussion

Pulse width can be either positive or negative. A pulse applied to the input of a monostable multivibrator causes it to switch state and stay in that state as long as the external timing network (*RC* circuit) is set. It then returns to its neutral state. The whole network is known as a "*one-shot*".

Pulse width is measured halfway between the upper and the lower edges of the pulse (overshoots and undershoots are not considered).

The external *RC* circuit should be connected to the designated pins of the device through a relay. This is a very sensitive test because of the stray capacitances that usually exist in the test hardware or cables; therefore, the resistor and capacitor used in the timing network must be carefully selected. The value of the capacitor plus the stray capacitance, the resistor, and the combination of the two must be within the pulse, or minus a few percent of the specified value (precision *R* and *C* may be necessary).

No TTL or MOS devices should be loaded in this test.

Definition

Pulse width (t_W) is the time between 50 percent of the upper and lower points on the output pulse of one-shot devices. This time must be performed with gross V_{IH} (nonthreshold input voltage).

Procedure

1. Connect the power supply to the power pin and 0 V to GND of the device.
2. Close relay(s) connecting the *RC* circuit.
3. Set appropriate V_{OH} and V_{OL} at the output(s).
4. Perform the measurement within the second half of the *RC* timing period.
5. The truth table sequence (functional patterns) must be applied to set the device in the one-shot condition.

3.19 Rise and Fall Time Test (t_R, t_F)

Definition

Rise time (t_R) is the required time for a pulse to complete its transition from a point at low level to a point at high level, usually from 10 percent to 90 percent of the output pulse.

Procedure

Refer to Sections 2.7.3 and 2.7.4 in Chapter 2.

1. Force V_{DD}/V_{CC} to the power pin and 0 V to the GND pin.
2. Specify a low and a high threshold voltage.
3. Connect the timing measurement system at each output.
4. Measure the time between for t_R.

The measured value must be between the suggested high and low limits.

All TTL and MOS devices must be loaded in this test.

3.20 Setup Time Test (t_{SU})

Definition

The setup time is the time between the application of a signal at a specified input and the next active transition at another defined input pin. Refer to the AC waveforms in Figure 3.11.

Note. Setup time is the actual time between events. It may be negative (compared to the *target* minimum or maximum limits) or too small to measure. A minimum value of this time is given by manufacturers that guarantees the logic operation of the device.

No TTL or MOS devices are loaded for this test.

Procedure

1. Apply V_{DD}/V_{CC} to the power pin and 0 V to the GND pin.
2. Set the timing generator with the appropriate cycle time. Set the delay between input data and the active transition edge (clock, enable, etc.).
3. Apply the proper pattern (from the device truth table) to the input(s) to maintain the device logic function and look at the output for the appropriate level and/or transition. The device must function correctly.

3.21 Hold Time Test (t_H)

Definition

This measures the time during which a signal is retained at a specified input after an active transition occurs at another known input terminal. Refer to the AC waveforms in Figure 3.11.

Note. Hold time is the actual time between two events and, like the setup time, may be too small to measure. A minimum value is usually suggested by the manufacturer that guarantees the logic operation of the device.

No TTL or MOS devices are loaded for this test.

Procedure

1. Force V_{DD}/V_{CC} to the power pin and 0 V to the GND pin.
2. Set the timing generator with the proper cycle time.
3. Do the functional testing at the desired pin.

3.22 Removal Time (t_{REM}) or Recovery Time (t_{REC}) Test

Definition

Removal and recovery time denote the same thing, which is the time required to change a memory type element from write to read mode condition while the predefined signal is still maintained at the output (definition from *The TTL Data Book*, 2d ed., Texas Instruments, Inc.).

No TTL or MOS devices are loaded in this test.

Procedure

1. Force V_{DD}/V_{CC} to the power pin and 0 V to the GND pin.
2. Set the appropriate signal and timing cycle to a timing generator.
3. Prepare the sequence of the truth table (see note).
4. Set delay between the input data and transition edge (clock, enable, etc.) to t_{REC} limit.

Note. When measuring t_{REC} from clear to clock pin, for instance, the clear and clock pins should stay in low condition for the first sequence of the logic pattern. In the next pattern, clear should go high and the clock must pulse. The triggering condition is dependent on the device type.

3.23 Noise in Digital Circuits

Since digital circuits are designed and implemented using a purely mathematical model, they may not operate within acceptable limits, due to noise. Noise is usually caused by radio frequencies (RF) and must be taken care of at the design stage.

The elimination of noise in digital and analog circuits is very different. One method of grounding suitable for noise elimination in an analog circuit may be a source of noise if implemented in a digital circuit. A small external noise introduced to an analog circuit can cause serious interference, especially in low-voltage operating circuits, while the same external noise is not a factor in the digital circuit. Therefore, external noise is of primary concern in analog circuits. Sources of external noise are ground lines, power lines, reflection from transmission lines, and crosstalk.

3.23.1 Ground Noise in Digital Circuits

The noise created in a digital circuit by an improper ground can be even more serious than the noise caused by the power supply. While supply source noise can often be suppressed by a decoupling capacitor, controlling and minimizing ground noise is more difficult.

A primary source of noise inside digital circuits is transient ground noise. Minimizing ground impedance is the best way (possibly the only way) in which the problem of transient ground noise may be resolved.

The inductance that every conductor possesses to some degree is also a significant problem in digital circuits, especially in circuits that operate at high frequencies. Inductance minimization at the design level is the answer to this type of ground noise.

In all conductors, the conductance has a logarithmic relation with the conductor's diameter. For the diameter d cm, h cm above the current path,

$$L = 0.005 \ln \frac{h}{d}$$

A conductor's inductance is independent of the shape of its cross-section as long as the surface area is the same. The inductance is slightly minimized by increasing the area of the cross-section.

Adding additional ground paths that are electrically parallel to the other ground paths is another way of minimizing inductance. This works because two electrically-parallel conductors have one-half the inductance of one alone. Similarly, four have one-fourth the inductance of a single path. Some mutual inductance, which increases the cumulative conductance, may be created between conductors that are very close together. Therefore, the parallel ground paths should be very close together. Any electronics text will provide the mathematical relationship between the inductors, distance, and mutual inductance. A digital system must have a solid noiseless ground, which can only be provided by proper design. A noisy ground can cause overwhelming problems that make the system unworkable. The best method of reducing ground noise at board level—a gridded ground plane—is not cost-effective. Locating the forward and return current paths close to one another, as in a twisted coaxial cable, can reduce the inductance of ground lines to some extent.

3.23.2 Reflection (Ringing)

Reflection, or ringing, is a transmission line effect that results in circuit noise and malfunction, especially in high-frequency operating circuits. CMOS circuits are immune to reflection due to their low speed and current requirements. Ringing can easily be seen on an oscilloscope, and cause a digital gate or flip-flop to trigger to a wrong state (refer to Figures 3.12 to 3.14).

Any conductor offers impedance to a current passing through it that is directly proportional to its length. If a voltage V_{in} is applied to a line conductor, a current V_{in}/Z_{line} is produced, which remains constant as long as the voltage does not change and the signal is propagated along the conductor. Impedance mismatch, the root cause of reflection, occurs when the conductor terminates to a device with either higher or lower impedance than the impedance of the conductor (Z_{line}). As reflection intensifies, propagation through the line decreases. If the end of the conductor terminates to a load or device whose impedance is equal to the impedance of the conductor (Z_{line}), there is no impedance mismatch and the signal will propagate through the line without reflection.

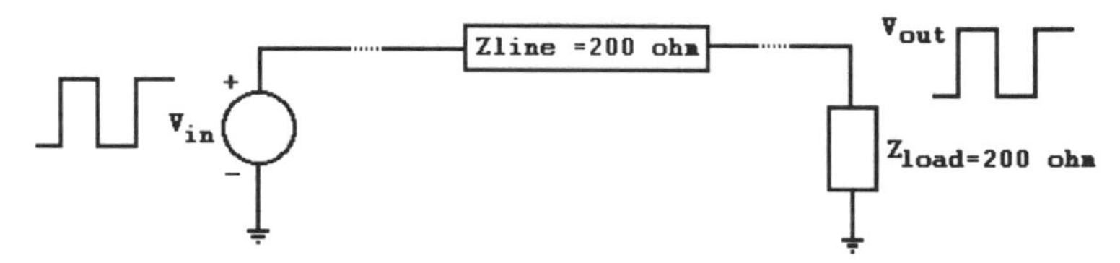

Figure 3.12 Case 1. Impedance match $Z_{line} = Z_{load}$: no reflection.

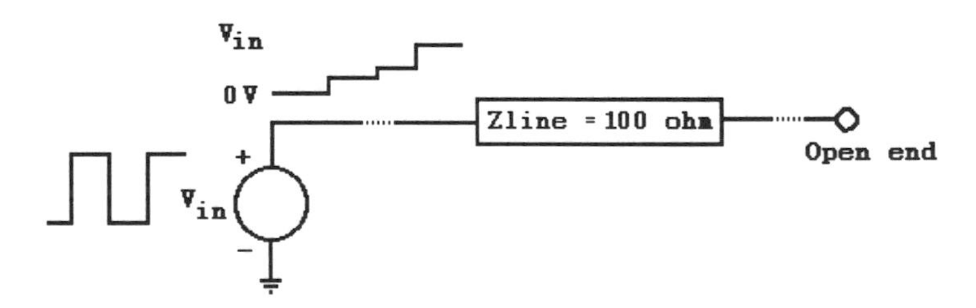

Figure 3.13 Case 2. $Z_{line} > Z_{load}$: ringing occurs at the sending end.

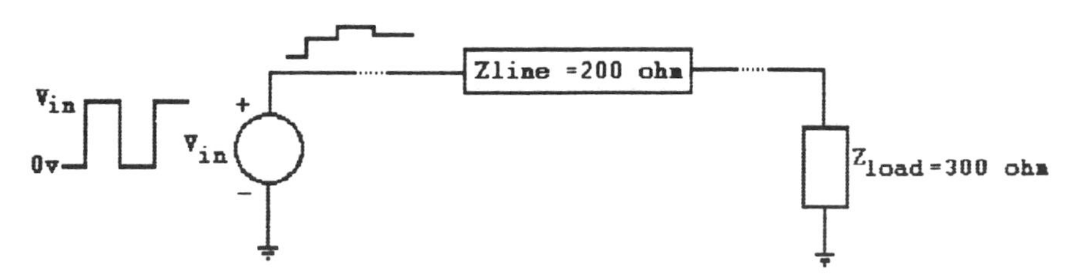

Figure 3.14 Case 3. $Z_{line} < Z_{load}$: reflection at both ends.

The length of conductors (buses) used in digital circuits can affect the rate of current change in the circuit and generate noise in digital networks. The conductor length between two gates creates a low transmission effect between the gates involved and induces capacitance to the network. The stray capacitance must be discharged before the next pulse is transmitted to the stage. The current discharged from the stray capacitance creates noise voltage due to the ground inductance and triggers gates to unwanted states.

The discharged current, combined with ground resistance, may create a resonance that can create oscillation. The resonance gain is given by

$$Q = \frac{1}{R}\sqrt{\frac{1}{C}}$$

The problem can be overcome by adding a small-value resistor to the output of the gate as shown in Figure 3.15. A reverse biased diode can also be used at the output of the gate, but this method is risky since high current will produce additional noise in the diode itself.

Stray capacitance, another source of noise, is illustrated in the totem pole configuration in Figure 3.16.

The reflection noise at the transistor level is described as follows. As they switch frome one state to another, transistors draw a large amount of transient current. This current is used to charge the stray capacitance as a load. Stray capacitance creates a momentary drop in the supply source as it passes through the supply and lines. This interruption causes a discontinuity that is reflection noise. Transistor Q2, in Figure 3.16, is Off and Q3 is On if a high output state exists; the reverse is true in a low output state. Both states create high impedance between power and the ground. Q2 and Q3 are both On for a very short instant

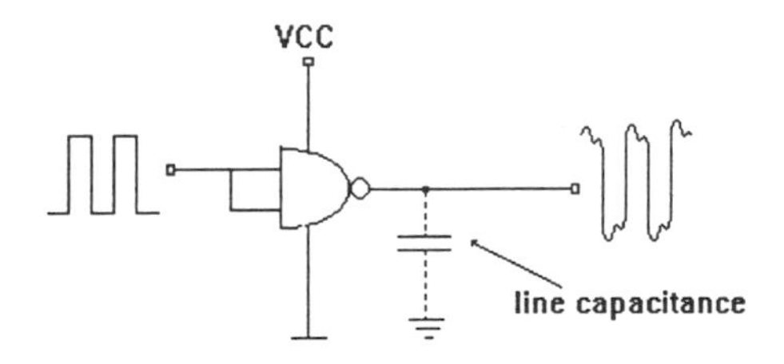

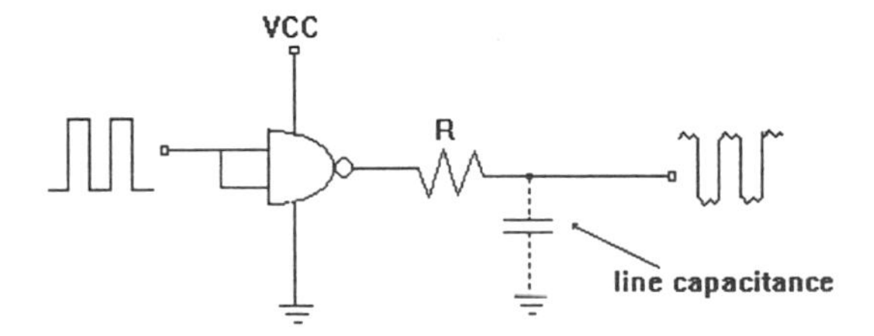

Figure 3.15 Effect of reflection on the output.

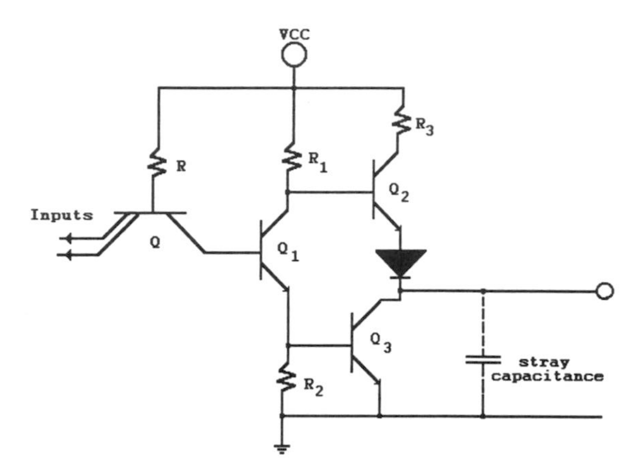

Figure 3.16 Totem pole circuit and stray capacitance at the output.

in the transition between saturation and cutoff, which causes the impedance to drop instantly and results in a current spike. Connecting a capacitor at each gate that supplies the transient current will eliminate the noise caused by this spike.

TTL is the only semiconductor technology in which mismatched impedance occurs. Depending on the logic operation of the device, there may be a large impedance difference between the inputs and outputs. There are a few simple, useful solutions to the problem of an ordinary impedance mismatch. Other remedies for reflection include:

- **Measurement strobing.** Insert some delay time between the application of the signal and the measurement. This helps to correct measurement strobing after the ringing period.
- **Overload.** If adding impedance to the end of the line causes overload, an added resistor must be connected to a voltage level within the V_{CC} and to a ground. This assumes that the impedance of the test equipment line receiver is much higher than that of the DUTs.

Thévenin's theorem is an appropriate tool to find the impedance of the other end of the line for impedance match purpose (refer to Figure 3.17). The resistor value that can be found this way is workable in most cases. The quickest method is trial and error, which in many instances saves engineering time. The theoretical approach to determination of the value of R_G is also presented.

On the basis of the DUT information provided by the manufacturer's data book, the following steps are required for finding R_C and R_G in Figure 3.17. In our case, the following information is assumed to have been provided:

$$V_{CC} = 5.0\text{ V}$$

$$I_{OL} = 18.0\text{ mA}$$

$$V_{IL} = 0.4\text{ V}$$

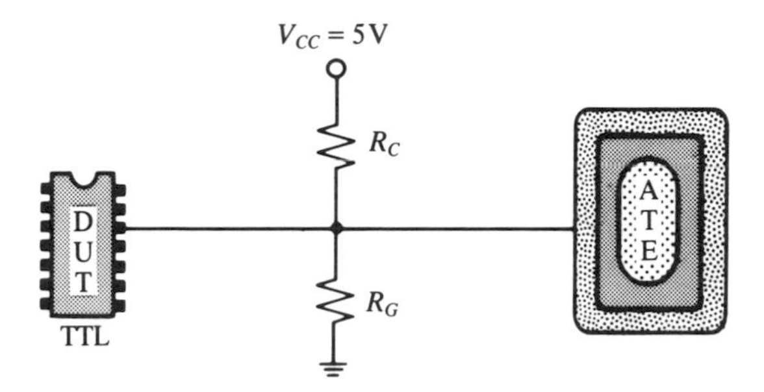

Figure 3.17 Finding the impedance of the other end of the line for impedance match purpose.

$$I_{OH} = -1.0\,\text{mA}$$

$$V_{IH} = 3.5\,\text{V}$$

The value of R_C is determined from the following simple relations:

$$\begin{aligned} R_C &= (V_{CC} - V_{IL})/I_{OL} \\ &= (5.0 - 0.4)\,\text{V}/18.0\,\text{mA} \\ &= 4.6\,\text{V}/18\,\text{mA} \\ &= 255\,\Omega \end{aligned}$$

The current through R_C is given by

$$\begin{aligned} I_C &= (V_{CC} - V_{IH})/R_C \\ &= (5.0 - 3.5)\,\text{V}/255\,\Omega \\ &= 1.5\,\text{V}/255\,\Omega \\ &= 5.88\,\text{mA} \end{aligned}$$

The value of R_G is then found from the above information.

$$\begin{aligned} R_G &= V_{IH}/(I_C - I_{OH}) \\ &= 3.5\,\text{V}/(5.88 + 1)\,\text{mA} \\ &= 3.5\,\text{V}/6.88\,\text{mA} \\ &= 508\,\Omega \end{aligned}$$

Therefore, the combination of R_C and R_G resistors can stop ringing to a great extent. Trial and error is probably the fastest method to find an approximate resistor value for this purpose. When the above methods are not applicable, the signal must be modified by implementing a buffer between the ATE and the DUT. This greatly decreases the possibility of any AC parametric measurement.

3.23.3 Crosstalk

Crosstalk is the noise created by magnetic or electric coupling between two close signal lines. The coupling is caused by mutual inductance and capacitance

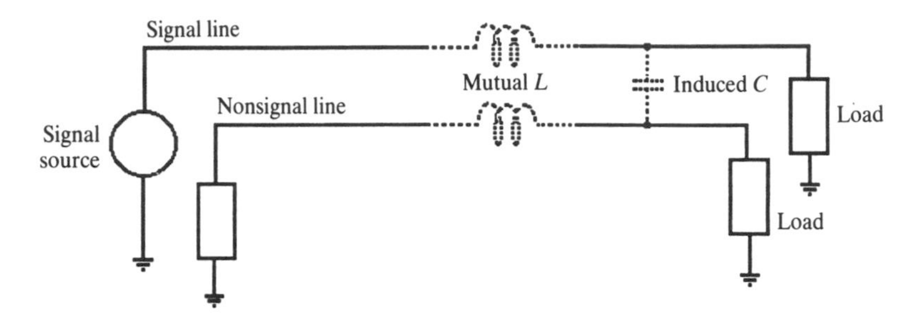

Figure 3.18 Crosstalk.

between the lines. The noise propagates in both directions, with the same polarity in the capacitive situation.

The instantaneous current changes that occur in high-speed logic circuits (described earlier and see Figure 3.16) can cause corresponding current spikes in quieter lines that are close to them. These spikes can cause an incorrect change in a logic gate by interfering with its threshold voltage level. Crosstalk is graphically depicted in Figure 3.18.

Crosstalk rarely exists in modern ATEs. It is a board-level design issue. If crosstalk noise is a manageable factor in testing, the following suggestions may remedy it.

- Use two conductors, one to carry signal and another from the DUT ground to the system ground.
- Use a coaxial conductor with a grounded shield.
- Twist together the signal-carrying conductor and the ground line.

References

1. Albert Paul Malvino, *Electronic Principles*, McGraw-Hill, 1984.
2. Arpad Barna and Dan I. Porat, *Integrated Circuit in Digital Electronics*, Wiley, 1987.
3. Daniel L. Metzger, *Electronic Circuit Behavior*, Prentice-Hall, 1983.
4. National Semiconductor Corp., *AS/ALS Logic Data Book,* 1990.
5. Texas Instruments, *ALS/AS Logic Data Book,* 1986.
6. J. A. Scarlett, *Transistor-Logic and its Interconnections, A Practical Guide to Microelectronic Circuits*, Van Nostrand Reinhold, 1972.
7. Texas Instruments, *High Speed CMOS Logic Data Book*, 1988.
8. Sandeep Patel, *Bipolar Logic Test Guidelines*, Standard Digital Logic Group, National Semiconductor Corporation, 1989.
9. Amir Afshar, *High Speed CMOS Logic Test Guidelines*, Standard Digital Logic Group, National Semiconductor Corporation, 1989

CHAPTER 4

Noise Identification

Introduction

Noise identification and elimination methods are extremely valuable for design, product, or test engineers at any level. In this chapter each noise is identified by its nature, effect, and cause, and illustrated with figures. Simple mathematical formulas and theoretical discussions are included, but the noise-suppressing methods are all practical and based on the experience of thc author and other materials of his suggestion.

Topics Definitions and sources of all types of electronic noise; best methods to identify and suppress noise.

4.1. Grounding, including single point, multiple point and hybrid grounding; general guidelines
4.2. General guidelines for grounding noise
4.3. Resistor noise—film type, wire wound, and combination type
4.4. Inductor noise
4.5. Capacitor noise, including electrolytic capacitors (polarized and nonpolarized); dipped mica; and ceramic
4.6. $1/f$ or flicker noise
4.7. Shot noise
4.8. Thermal noise
4.9. Popcorn (burst) noise
4.10. Contact noise
4.11. Using capacitors for suppressing noise
4.12. Decoupling
4.13. Facts about power supplies, including *RC* and *LC* decoupling
4.14. Suppressing noise created by capacitive load
4.15. Noise in bipolar transistors

The undesirable signals present in electronic or electric circuits are referred to as noise. Noise will degrade the performance and efficiency of the circuit if it is not controlled or eliminated. Because of the many factors involved in the creation of noise, it is critical to know the cause before deciding how to eliminate it.

Discontinuity of electrical charges is the primary problem related to noise. These charges constitute the noise, or unwanted signal, when carried as pulses in discrete quantities in a conductor. They become more consequential as they are multiplied by the amplification stages. Some electronic devices are vulnerable to environmental noise because of the nature of their internal structure. These devices pick up noise from their surroundings that, even in small amounts, degrades their operation. The noise generated by the nonlinearity of a circuit is a design problem and does not concern test engineers.

Although noise cannot be totally eliminated from a circuit, the best solution is to track it down to the source and properly suppress it there. The three major steps in eliminating noise are

1. Identifying it.
2. Isolating noise-sensitive devices from noise sources.
3. Eliminating the path carrying noise to the circuit or the sensitive element.

Sometimes more than one method of suppression is needed to achieve an acceptable result.

This chapter concentrates on providing test engineers with a quick and effective way of identifying and eliminating noise. The complex mathematical equations, such as Maxwell's three-dimensional space coordinate system, for noise detection or reduction are not included. In many instances, the calculated values of components or variables found this way are useless in a practical sense because of the presence of many uncontrollable variables and hard-to-define boundary conditions. Most books on noise are based on mathematical theories and are difficult to follow because of their complex mathematical methodology. The most practical source used, a valuable book by Henry W. Ott of Bell Laboratories, is *Noise Reduction Techniques in Electronic Systems.*

Test engineers do not usually employ theory when dealing with aspects of noise. They prefer to detect the sources of spikes and glitches and to eliminate the side-effects without using calculus and sophisticated calculations. This chapter contains the essence of all the different topics, textbooks, practical test engineering class notes, and on-the-job experience that have been accumulated by the author during his years as a test engineer.

The most fundamental step in noise reduction is the precautionary and preventive consideration of noise during the design stage of any circuit. A test engineer deals with a test system, a device under test (DUT), and test hardware, any or all of which can be the source of noise. Some noise, such as that generated by equipment (the power supply or the signal generator) or inherent noise in the device, is outside the test engineer's control. All other noise being generated by the interfacing hardware and cables must be identified and controlled or eliminated. Current or signal-carrying conductors should be located away from any environment containing magnetic fields or any type of radiation. The major factors involved in preventing noise and eliminating its effects are discussed in the following pages.

4.1 Grounding

Grounding, if it is done properly, can be the primary answer to noise problems. Certain facts about grounding must be known before a good grounding system can be provided. Methods of signal grounding may vary from being all at the same location, being at different sites, or a combination of both (hybrid). Every conductor has its own impedance plus inductance limitations; also, individual grounds that are separately connected to different spots do not have the same electrical potential. Each manner of connection provides appropriate safety and has a specific effect on noise reduction.

Methods of signal grounding are categorized into three different groups. Each group is described below.

1. Single-point grounding. Single-point grounding has the advantage of being the simplest method of grounding, but it is not very effective for controlling noise. When two systems are operating with a wide range of voltage level, it is particularly ineffective. Figure 4.1 illustrates single-point grounding. The voltage drop across each of the conductors in this figure is different because each conductor has its own finite impedance. Because of this difference, the current, and thus the nature of the ground (determined by current) will be different. The current in each branch is shown in the figure.

This method avoids adding the impedance of the path connecting the DUT to the ground to the next point. Unfavorable aspects of single-point grounding are:

- As the frequency increases, the ground impedance also increases.
- With increases in frequency, the conductors radiate noise to the surrounding parts of the circuit and may create crosstalk to other quiet neighboring buses.

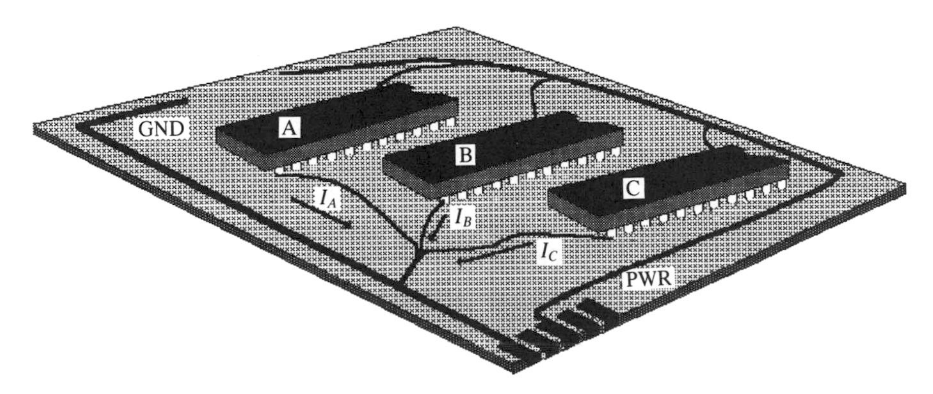

Figure 4.1 Single-point grounding suitable for low frequency.

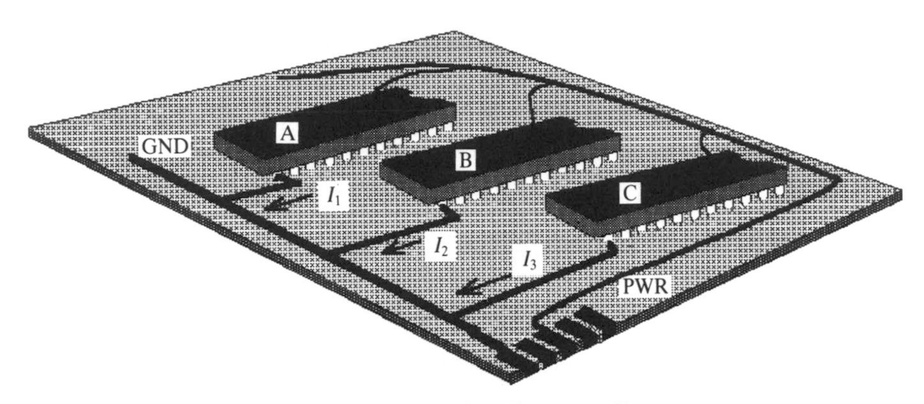

Figure 4.2 Multiple-point grounding.

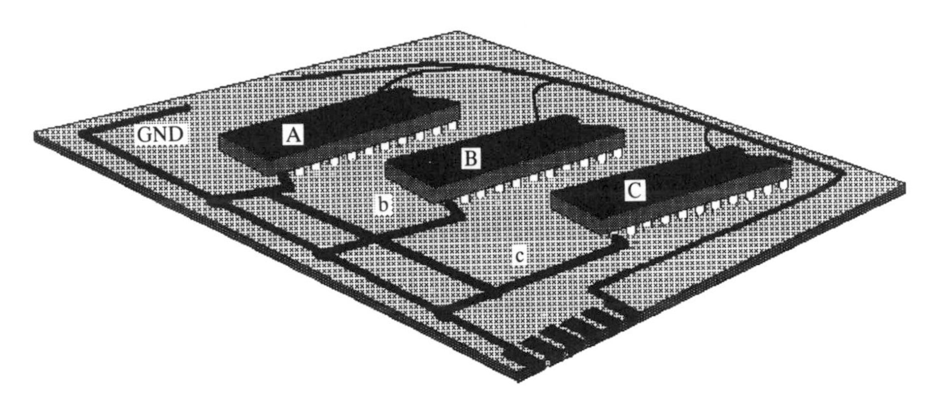

Figure 4.3 Hybrid grounding suitable for high or low frequency.

An important concept in this type of grounding is that the conductor that connects the circuit to the ground must be as short as possible to eliminate impedance.

2. Multiple-point grounding. In multiple-point grounding, each separate system is individually connected to a spot on the ground line, and thus it is an ideal method for use with high-frequency digital circuits. This method of connection creates very low ground impedance that can be further reduced by making all the circuit-to-ground connections as short as possible. Figure 4.2 shows a type of multipoint grounding system.

Multiple-point grounding is not desirable for low-frequency applications because it causes all the system current to flow through the ground plane. The expansion of the ground plane by any method available will not upgrade the circuit performance in low-frequency applications.

3. Hybrid grounding. This method is a combination of the two methods described earlier. The performance of hybrid grounding systems varies with changes in frequency. The behavior changes from single-point grounding to hybrid as the frequency increases. Figure 4.3 shows a schematic of this type of grounding.

Grounding can become a source of noise under the following conditions.

1. When the circuit operates with low-level analog signals.
2. When ground points are terminated to an AC power ground or are too far from each other.

Reducing the type of grounding to a single-point method can solve problem (2). In the case of an op amp, shielding, especially for high-gain amplifiers, can be a solution. Whether or not the common point is grounded, a shield or coaxial cable should be grounded to the common terminal of the amplifier to suppress noise created by the cable resistance.

A shield should be grounded at both ends when it is longer than 1/20 of the wavelength or when the noise from the passage of power frequency appears in the shield. This last noise occurs at low frequencies of less than 1 MHz. Grounding the shield at both ends creates a magnetic shield at high frequency and is very useful in digital circuits with high-frequency signals.

There are elements that are capable of producing noise in any network or device, so that the signal-to-noise ratio of the network is degraded when these elements are activated. A change in the load resistance in a terminal will change the noise by the same factor; therefore, the signal-to-noise ratio for any network with the same frequency band will not change.

Although semiconductor devices are designed for particular purposes and operations, the technology employed in their fabrication categorizes them with respect to their potential noise creation. For example, the nature of noise in bipolar technology is not the same as in MOS. In bipolar transistors, the primary type of noise is shot noise (which will be discussed in Section 4.7). In MOS transistors, transient spikes occur when power is applied to the device and they are very sensitive to capacitive loads. Although no component is perfect, resistors, capacitors, inductors, and most common components are normally treated as perfect in theoretical calculations.

Because of their physical shape or internal structure, any of the components mentioned above will contribute some noise to a circuit, and this noise can sometimes be very serious. For designers and test engineers, understanding the behavior of passive elements used extensively in electronic circuits is essential for their proper use. These components, used singly or in combination, can be a new source of noise and can further complicate the noise reduction process.

4.2 General Guidelines for Grounding

The following are general guidelines for grounding noise from common sources.

Noise Source	**Grounding**
Signal hardware and noisy grounds	Use a single-point grounding system with more than a few separate returns.
> 1 MHz frequency	Use multiple-point grounding.

Noise Source	Grounding
Amplifier grounded/signal source not grounded	Ground the cable shield at the common point of the amplifier.
Signal source grounded/amplifier not grounded	Ground the cable shield to the common point of the signal source.
Amplifier protected with shield	Ground the shield to the common point of the amplifier.

4.3 Resistor Noise

Resistors can be classified into three distinct categories according to the processes and materials used to manufacture them:

1. Film type resistors
2. Wire-wound resistors
3. Combination type resistors

Each of these three types of resistors has some capacitance that varies slightly from the others.

Of the three types of resistors, the wire-wound is the quietest and the combination is the noisiest; both generate primarily thermal noise. The film type resistor is quieter than the combination type and noisier than the wire-wound.

The amount of noise in a resistor depends mainly on the shape of the component and the type of material in its structure. The resistor lead is the source of inductance and makes the element vulnerable to picking up surrounding magnetic fields, and the inherent capacitance of any resistor is about 0.5 pF. As the resistor value increases, this capacitance will be more effective. As the frequency increases, the capacitive reactance will become a major part of the resistor value and may degrade the circuit operation.

Contact noise is an additional noise source in the combination type resistors discussed in Section 4.10. These types of resistors generate contact noise when receiving current as a result of their different element structure.

The wiper contact of the variable (adjustable) type resistor (occasionally used on boards) is one more source of noise that is directly proportional to the current and the value of the selected resistance.

A resistor produces noise voltage in an inverse relationship with its power rating. The higher the power rating, the quieter is the resistor with the same resistance value.

Different resistor types are chosen for different applications. For highest precision, the best choice is generally the wire-wound type. The bifilar wire-wound type is especially desirable for high frequency, but the inductance of the resistor becomes an adverse factor.

The equivalent circuit for a resistor is shown in Figure 4.4.

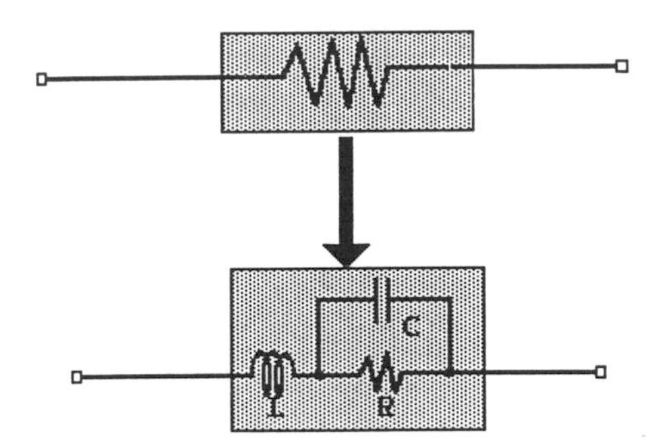

Figure 4.4 Equivalent circuit for a resistor.

4.4 Inductor Noise

Every inductor has some resistance because it is wire-wound and has capacitance between the wire layers. Theoretically it should have only inductance. An equivalent schematic for an inductor is given in Figure 4.5.

Every inductor has a wire wound around a core. Inductors are categorized into two groups based on the type of core:

1. Magnetic core
2. Air core

The magnetic-core inductor, either open or closed, is the more susceptible to magnetic fields. The open core type has more problems with generating magnetic fields than any other inductor, although the problem is inherent in inductors.

A low-resistance material such as aluminum or copper can be used to shield the magnetic field and to prevent the passage of magnetic flux at higher frequencies. At very high-frequency operation, inductors are safe from noise. Inductance, resistance, and the quality factor (Q) are critical factors in any inductor. In the magnetic core type, the Q is a result of the first two factors together with the

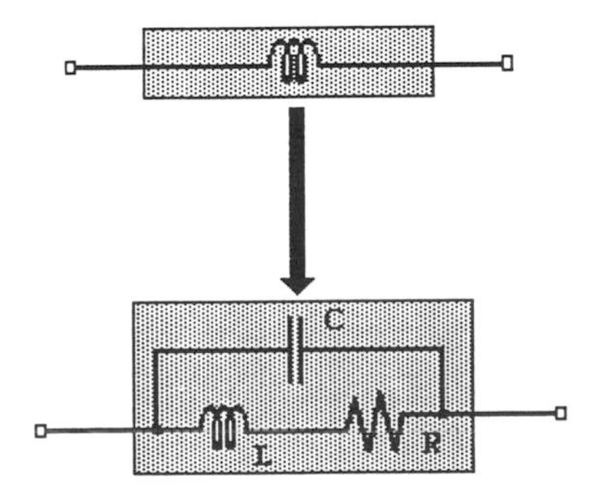

Figure 4.5 Equivalent circuit for an inductor.

maximum DC rating. Quality factor is the ratio of coil resistance to DC resistance, given by

$$Q=\frac{XL}{R_{\mathrm{dc}}}=\frac{\omega L}{R_{\mathrm{dc}}}$$

where ω is the frequency in hertz, X is the inductive reactance, L is the inductance in henrys, and R is the resistance in ohms.

Note. The inductive and capacitive reactance are calculated from the following formulas respectively:

$X_L = j\omega L$

$X_C = -j/\omega C$

where $j=\sqrt{-1}$, the known imaginary number, C is the capacitance in farads, and L is inductance in henrys. For instance, the capacitive reactance of a capacitor of 1 μF at MHz frequency is $X_C = -0.16j$ ohms.

4.5 Capacitor Noise

All capacitors have two material electrodes spaced by an insulator called a *dielectric*. The basic characteristics of a capacitor are capacitance, frequency response, voltage rating, dissipation factor, insulation resistance, and sometimes conformation and size; some of these characteristics are temperature-sensitive. The combination of these parameters does not result in a pure capacitance element. The capacitance in an ordinary capacitor is associated with traces of resistance and inductance, as shown in Figure 4.6. The inductance L originates from the device lead and R is due to the resistivity of the dielectric (insulator) material that separates the electrodes.

The capacitance is given by

$$C=\frac{KA\varepsilon_0}{d}$$

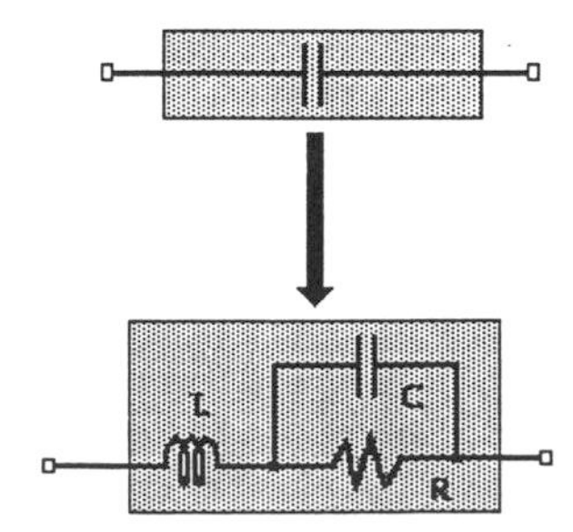

Figure 4.6 Equivalent circuit for a capacitor.

where

K = dielectric constant of the insulator used
A = the surface area of electrodes in m^2
ε_0 = permittivity of free space (constant)
d = electrode separation in m

The dielectric constant primarily determines the sensitivity of any capacitor to frequency and temperature. The most crucial capacitor characteristic affected by the capacitor inductance and its load is its frequency response. This inductance characteristic causes the capacitor to become self-resonant at a frequency that exceeds its resonance limits. Inductive reactance, which causes impedance to surge as the frequency increases, is also created by inherent inductance.

There is some power loss in the form of RI^2 in any capacitor, which occurs in the dielectric material as a result of the random motion of particles in the dielectric material (insulator). A single capacitor is not able to cover all ranges of frequencies but, when filtration of different frequencies is desired, capacitors can be used in parallel. Paralleling capacitors can, of itself, produce a resonance problem that results in maximum impedance at certain frequency ranges. This problem occurs most often when very different capacitor values are used in parallel.

Capacitors are grouped into the following types:

- Ceramic
- Film type
- Mica
- Electrolytic
- Trimpot

The choice of one group over another is dependent on the nature of the application. Of the types named here, the electrolytic polarized type best lends itself to further discussion.

4.5.1 Electrolytic (Polarized) Capacitors

Besides their higher voltage-handling capability, the prime advantage of electrolytic capacitors is that they can be produced in small sizes with large capacitative values of up to 100 mF. Their main disadvantage is that they frequently explode because DC polarity must be maintained to prevent excessive current leakage. Consequently, the electrolytic capacitor should not be operated at its maximum voltage-handling capability. A nonpolarized capacitor is created when two electrolytic capacitors of the same value are connected together in series so that they are poled on opposite sides.

4.5.2 Electrolytic (Nonpolarized) Capacitors

Electrolytic nonpolarized capacitors are useful for low-frequency bypassing, buffering, and decoupling (a minute amount of current crosses this device as leakage). They come in relatively large capacitance values (up to 20 μF) and have better low-frequency response than the polarized type. Applications for electrolytic capacitors are discussed in Sections 4.13.1 and 4.13.2.

4.5.3 Dipped Mica Capacitors

Dipped mica capacitors are accurate and moisture-resistant: They are suitable for load, excellent for high-frequency applications, and are available in small values; they are, however, relatively large in size.

4.5.4 Ceramic Capacitors

Ceramic capacitors are superior for bypassing high-frequency supplies. They are available in small physical sizes and low capacitive values.

4.6 1/f or Flicker Noise

Most devices are capable of generating flicker noise, which is directly dependent on current flow. The main source of this noise is contamination at the emitter/base layer of the crystal in transistors and diodes. Flicker noise is worse when frequency decreases: it has a $1/f$ relationship with frequency.

The general relation for $1/f$ noise current and its dependency on frequency is given by

$$I = \sqrt{\frac{KI\,\Delta f}{f}}$$

where

Δf = bandwidth of a portion of frequency in hertz
I = DC current in amps
K = a constant that can be approximated and varies from device to device and even from one transistor to another on the same die

The average value taken from many transistors on the same wafer will yield the best value for K. Thus K is directly affected by imperfection and contamination in the wafer process.

4.7 Shot Noise

The majority charge carriers crossing the barrier of a metal semiconductor in either direction produce this type of noise. This barrier causes a fluctuation in the current flow called *shot* or *Schottky noise* that comes from the random motion of electrons or holes. Accordingly, this type of noise is directed by current flow and is common in diodes and bipolar transistors, and its characteristics are similar to those of thermal noise.

The following relationship holds for Schottky noise current that is independent of frequency:

$$I_{shot} = \sqrt{2I_{dc} q \,\Delta f}$$

where

I_{dc} = average DC current in amps
Δf = bandwidth of noise in hertz
q = electron charge (1.6×10^{-19} coulombs)

This type of noise is totally independent of temperature.

The magnitude of this noise can be determined by measuring the value of current passing through the device, which is directly proportional to bandwidth.

4.8 Thermal Noise

Thermal noise (also known as Johnson noise) is created in any component having some resistance by the excitation of electrons due to heat. The other name for this type of noise is *resistance* or *white noise*. It varies positively with temperature and the following relationship shows its dependency on other factors:

$$V_{th} = \sqrt{4T \,\Delta f \,Rk}$$

where

T = temperature in kelvin
Δf = bandwidth of noise in hertz
R = resistance in ohms
k = Boltzman constant (1.33×10^{-23} joules)

The noise power is independent of resistance for any specified bandwidth:

$$P_{th} = kT \,\Delta f$$

Noise current, unlike noise voltage, should be read with a low-impedance meter. The two relations above deal with thermal noise produced in an ideal resistor. In reality, there are other elements in the structure of resistors in very small amounts that create noise that may not be significant here. In noise-sensitive circuits it is best to avoid carbon-composition resistors, which are noisier than other types.

4.9 Popcorn (Burst) Noise

Some semiconductor diodes or circuits produce noise called *popcorn* (because it sounds like popping corn when it is amplified and heard through a speaker) or *burst*. Popcorn noise occurs when the base/collector junction of a transistor presents negative resistance at the knee of the current/voltage characteristic curve. It is a process-related type of noise and is more common in P-N-P than N-P-N type transistors. High-impedance circuits will have the highest popcorn voltage noise. Impurity of the metal junction in a semiconductor, which is the primary cause of this noise, is correctable if the impurities are removed during the fabrication process.

In a forward biased junction, the characteristic curve of popcorn noise is different from that of one that is reverse biased. Stress on the die on which the transistor is implanted and the emitter current are responsible for the repetition of popcorn noise. The popcorn current noise and its relation with other factors is given by

$$I^2 = \frac{KI\,\Delta f}{1+\left(\dfrac{f_1}{f_2}\right)^2}$$

where

I = DC current in amps
K = device-dependent constant (ranges from 0.5 to 2)
Δf = bandwidth portion of noise in hertz
$f_1 f_2$ = noise frequency on either side of the band

4.10 Contact Noise

Contact noise is caused by the nonuniformity of electron density in a conductor. When current passes through a nonuniform conductor (solder joints, switches, relays, etc.), it creates contact noise, which is another kind of low-frequency noise.

The relationship between the noise current and frequency is

$$I_{\text{noise}} = \frac{KI_{\text{dc}}\sqrt{B}}{\sqrt{f}}$$

where

I_{dc} = average DC current in amps
B = frequency bandwidth of the noise, in hertz
f = frequency in hertz
K = constant related to the conductor material and shape

Contact noise is a low-frequency-related noise and decreases with the increasing frequency as shown in the above formula. This noise is also called $1/f$ by some authors.

There is no single clear-cut remedy for all types of noise, since each noise is unique in nature depending on its sources. Most types of circuit noise can be radically suppressed by one, or both, of the following methods:

- **Thermal noise.** Decreasing the ambient temperature reduces the noise that results from the random motion of electrons in a conductor. This is the most effective remedy for thermal noise, especially in resistors.
- **Shot noise.** Lowering the current passing through devices, such as amplifier and diodes, can significantly reduce shot noise.

Section 4.11 describes the use of capacitors in noise suppression.

4.11 Using Capacitors for Suppressing Noise

Some straightforward clues to identifying and suppressing a noise source to such a degree that it does not affect the circuit or system performance were given in Chapter 3. Capacitors are the most useful of all available elements for dealing with all types of noise, especially when frequency is a major contributing factor. Some facts about capacitors must be appreciated before they can be used effectively. Some of the most significant of these facts follow.

- **Minimum finite impedance.** At a certain frequency, every capacitor possesses a minimum finite impedance.
- **Impedance degradation.** When capacitors of different values are used in parallel, the total impedance for all the frequencies involved will be significantly lowered.
- **Lead length.** In order to reduce inductance and resistance as much as possible, the length of the capacitor lead must be as short as possible.
- **Resonant frequency.** A capacitor's resonant frequency is inversely proportional to the value of its capacitance. In other words, the lower the frequency, the higher the capacitance.
- **Low-frequency requirements.** Low-frcquency applications call for nonpolarized capacitors.
- **Connecting capacitors.** Each capacitor should be connected to a different point on the ground line when covering several frequency spectra and using multiple capacitors.
- **Leakage.** There is always some leakage across any capacitor, no matter how small the capacitance is. To prevent leakage from causing problems in other DC measurements, it is appropriate to use a rclay to connect the capacitor to the ground point when it is not needed.

4.12 Decoupling

Figure 4.7 shows a method of decoupling different frequencies of a supply source using different component values. A bulk capacitor can prevent oscillation in the supply source and reduce low- to high-frequency noise. Fewer low-value capacitors can be used if there are restrictions.

Use of capacitors on the input or output of a device pin has one major disadvantage—its effect on the dynamic operation of the device. This effect is especially marked on the rise time. The rise time will be prolonged by the capacitor chargeup time, which is proportional to current. The relation between capacitance (C), voltage (V) and current (I) is

$$I = C\frac{\Delta V}{\Delta t}$$

where (t) is the time.

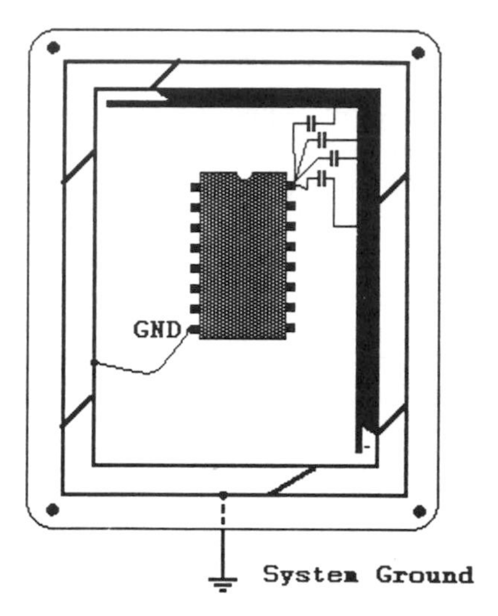

Figure 4.7 A good method for connecting different capacitors to GND path.

When the junction of a transistor is forward biased, a CMOS device has a tendency to draw excessive current. This condition is called *latch-up*. In latch-up the V_{CC} and the ground are shorted together. Several sources can provide the excessive current and create this condition. Improper sequencing of power sources, in case there is more than one supply source, and overpowering the device are two of the causes. In CMOS device testing, this condition can be prevented by using capacitors of the same value on all power supplies. Also, the proper voltage should be applied to the V_{CC} pin. Inadequate voltage may cause interference between pins and crosstalk.

It is better to use more than one individual ground return on circuits that are using varying power. The ground of the circuit that has a low power level should be separated from the one with higher power.

4.13 Facts about Power Supplies

A power supply that has zero impedance is ideal, but no such ideal supply exists. Therefore, each circuit design carries its own finite basic impedance that is the result of internal inductance and capacitance as shown in the following relation:

$$Z_C = \sqrt{\frac{L}{C}}$$

If a power supply is shared among several circuits, it should be capable of supplying constant voltage and current to all of the different loads connected to it.

Loads can disturb the constant flow of the output of a voltage supply and the switching activities of a digital device can be propagated to other devices through the power supply and its buses. This interference is of special concern since high noise levels can be produced in synchronous systems where devices switch simultaneously between ground and power lines. The power source itself and its decoupling capacitor, along with the conductor, offer some resistance and capacitance to the circuit.

Since the contribution of the effects of these elements to noise creation are significant, a variety of techniques is required to design a quiet network. Different techniques for suppressing noise in every stage of the circuit, when the power supply and its distribution network are under the test engineer's control, are offered here.

4.13.1 RC Decoupling

There are many ways of decoupling by *RC* network that can isolate and minimize the noise coupling through the supply source and prevent its spread, say, to circuit A as shown in Figure 4.8. Additional capacitors may be added to the circuit for extra filtration.

4.13.2 LC Decoupling

LC decoupling is ideal at low frequencies if the resonance frequency of the *LC* decoupling used at the power supply source is not an adverse factor. *LC* decoupling is highly recommended for power supply decoupling at high frequencies. The resonance frequency of an *LC* network is given by

$$F_r = \frac{1}{2\pi\sqrt{LC}}$$

The bandpass of the circuit that an *LC* network is used for must be higher than the *LC* resonance frequency. The inductor used in this filtering network must be able to pass the required DC current to the curcuit without becoming saturated.

If the rate of damping is not adequate, an additional resistor may be connected in series with the inductor. A capacitor, C_2 as shown in Figure 4.9, can be added to the circuit to further prevent noise affecting circuit A.

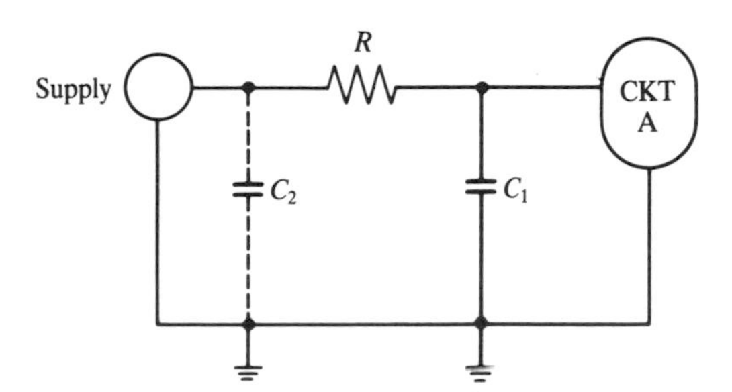

Figure 4.8 Power supply *RC* decoupling.

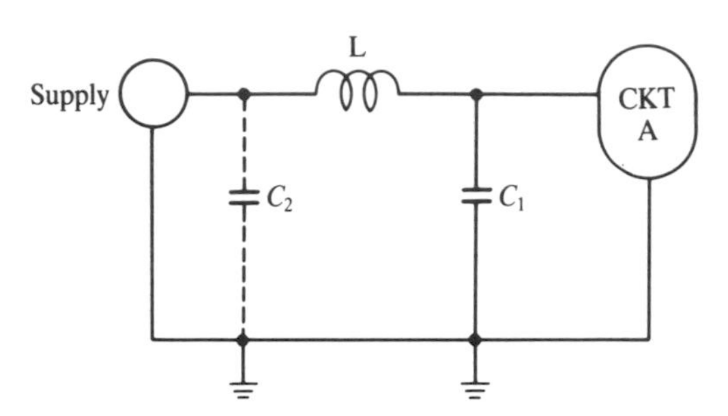

Figure 4.9 Power supply decoupling using an *LC* circuit.

4.14 Suppressing Noise Created by Capacitive Load

Oscillation at high frequencies brought about by inadequate power supply decoupling can occur when a large capacitative load, such as a long transmission wire, is connected to the emitter of an emitter-follower configuration. As the frequency increases:

- The collector impedance can increase due to the inductance of the power source.
- The line capacitance and the frequency cause the emitter impedance to decrease.
- Feedback around the transistor through the resistor is created when the transistor has a large voltage gain. This feedback is a common cause of oscillation that can be prevented by connecting a capacitor (with a value greater than the capacitive load of the transmission line) between the power supply pin and a high-frequency ground point of the amplifier circuit, as shown in Figure 4.10.

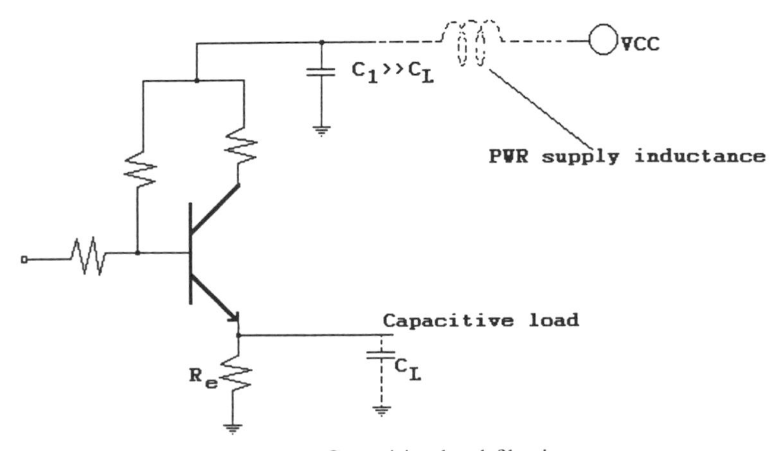

Figure 4.10 Capacitive load filtering.

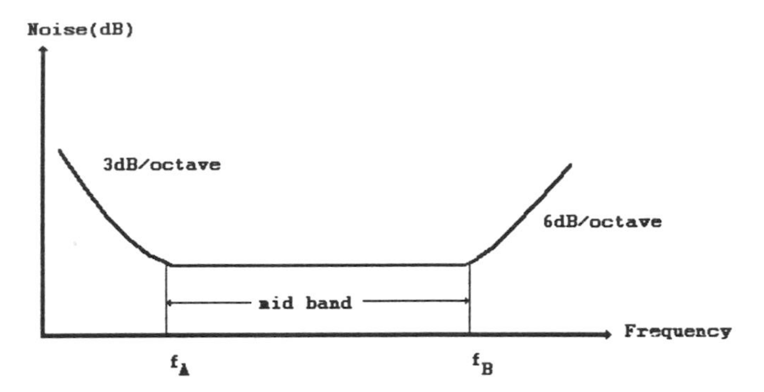

Figure 4.11 Noise vs. frequency in a bipolar transistor.

4.15 Noise in Bipolar Transistors

There are many types of frequency-dependent noise in bipolar transistors. When the transistor is in forward-active mode, shot noise is produced by the random motion of minority carriers crossing the base, arriving at the collector region, and producing random current pulses in that region.

The charge carriers extending from the base into the emitter region produce similar shot noise in the base current. Thermal noise is created by the base and the collector resistors in the network, and flicker and burst noise are generated at the base/emitter junction. Bipolar noise versus frequency is shown in Figure 4.11.

- *Above frequency f_A*: Thermal noise comes from the base resistor. Flicker or burst noise is produced at the transistor's collector and emitter junction.
- *Above frequency f_B*: There is an inverse relationship between noise produced at the collector junction and the transistor gain as the frequency increases ($>$ 10 MHz).

References

1. D. A. Bell, *Noise and the Solid State*, Wiley, 1985.
2. Henry W. Ott, *Noise Reduction Techniques in Electronic Systems*, AT&T Bell Laboratories, Wiley, 1976.
3. M. J. Buckingham, *Noise In Electric Devices and Systems*, Wiley, 1983.
4. David A. Bell, *Fundamentals of Electronic Devices*, Reston Publishing Co., 1975.
5. *Decoupling Your Way to Noise Reduction*, PIER Electronics, Inc., 1988.

CHAPTER 5

Operational Amplifier

Introduction

This chapter is devoted strictly to the operational amplifier, or op amp. No analog circuit is better known than the operational amplifier, which is a true analog device and is used in virtually any analog network. Transistors that construct this device, unlike those in digital devices, operate in the region between the cut-off and saturation.

We begin with the introduction of a simple op amp and its internal structure with in-depth description of the function of each internal component. We then move on to identify and explain a suitable method to test each parameter relying on the most advanced and accurate way of parameter measurement.

Every test procedure parameter is identified and explained thoroughly with the least amount of mathematics and theory possible. The sources of input offset voltage (VOS), input offset current (IOS), and slew rate (SR), and all other op amp characteristics, are demonstrated by schematics and explained in simple terms.

Aims

- Basic structure of an op amp.
- Relationship of each transistor with another and the voltage current.
- Definition, intention, and step-by-step procedures for op amp test using a nulling circuit. The position of every relay in the nulling circuit is shown clearly.
- Identification of sources of oscillation and other types of op amp noise and methods to suppress each type—clearly described in words and with graphics.

Topics

5.1. Basic structure of an op amp, and its ideal behavior
5.2. Feedback configurations
5.3. Basic op amp internal structure, including tail current, common mode

operation, offset voltage; analysis of basic op amp (loop gain and slew rate); inverting and noninverting amplifiers

5.4. Common mode signal with discussion of CMRR
5.5. Input impedance and output circuit
5.6. Single-ended input
5.7. Input bias current (I_B)
5.8. Slew rate, ideal response by op amp, sources of, test for
5:9. Op amp DC measurement (nulling circuit) with the position of every relay in the nulling circuit shown
5.10. VOS test, including causes of VOS, methods to diminish, measurement of
5.11. VOS test using nulling circuit
5.12. IOS test (input offset current)
5.13. VON test (negative output voltage swing)
5.14. VOP test (positive output voltage swing)
5.15. CMRR test (common mode rejection ratio)
5.16. SR test (slew rate)
5.17. AV (gain) test
5.18. PSRR test (power supply rejection ratio)
5.19. Noise in op-amp circuits, discussion of units for measuring
5.20. Oscillation, including feedback, causes, manufacturer's and author's suggestions for compensation
5.21. Noise in different types of op-amps (bipolar, FET), reducing op-amp noise through voltage divider, using capacitance, reducing DC offset and op-amp drift
5.22. Decibel, definition and equations used for

General Information

Analog vs. Digital Circuits

In analog circuits the voltage and current waveforms are similar to the signal variation, while in digital circuits there is a group of binary signals. These binary signals can be considered as High or Low (On or Off) based on the signal amplitude.

Digital pulses can be transmitted more efficiently and precisely than analog pulses because noise does not degrade these pulses as it does analog signals. In addition, digital readouts are much easier to read than the corresponding analog readout.

In testing digital circuitry we usually deal with either high or low state of the input or output, while measuring an analog device involves an infinite number of levels between low (ground) and high. Each level can have a decisive effect in the overall circuit operation. Any voltage that is added to or subtracted from an existing analog voltage creates an error that is added to or subtracted from the measurement system. This important aspect of analog pulses will become more evident when dealing with conversion of an analog pulse to its digital equivalent or vice versa (refer to Chapter 6 on A/D conversion). As a perfect example of an analog circuit we have chosen an operational amplifier, which is widely used in almost any electronic circuit.

An operational amplifier (op amp) amplifies, or enlarges, the input signal. At the transistor level, application of an AC voltage signal at the emitter of a transistor that is correctly biased in its active region causes a fluctuation in the collector current. This current then produces an output voltage larger than the input when it passes through an external resistor.

An operational amplifier is described here in terms of its ideal behavior. In the real world an ideal op amp does not exist because of all of the uncontrollable parameters. As the discussion proceeds, you will be able to understand the source of any undesirable symptoms that may arise in the device test debugging process. To clarify all aspects of an op amp, a schematic of the internal structure of a basic op amp and a description of its elements and their functions are included.

Commercially available op amps have two inputs, marked (−) and (+) to show the negative (inverting) and positive (noninverting) inputs, respectively, and an output terminal. If a signal is applied to the (+) input, it will have an output that is *in phase* (has a positive gain) with the applied signal. The output of a signal applied at the (−) input will be *inverted* (has a negative gain).

5.1 Ideal Behavior of an Op Amp

Ideally an operational amplifier would exhibit the following characteristics:

- Zero DC output voltage when the inputs are connected to ground level.
- On the basis of the following relation, the op amp can be considered as a linear voltage-controlled voltage source:

$$V_{out} = -AV_{dif}$$

where A is the gain of the device and V_{dif} is the differential input voltage.

- Since the impedance between the two differential inputs is infinite, any feedback signal has no effect on the impedance and, therefore, no input current exists.
- Infinite open loop voltage gain.
- Zero propagation delay.

Since we have already learned that uncontrollable parameters make ideal behavior impossible, in practice an op amp behavior should exhibit the following:

- Extremely high frequency gain
- Extremely high DC gain
- Very low output impedance in closed loop
- Uniform gain covering a wide frequency spectrum

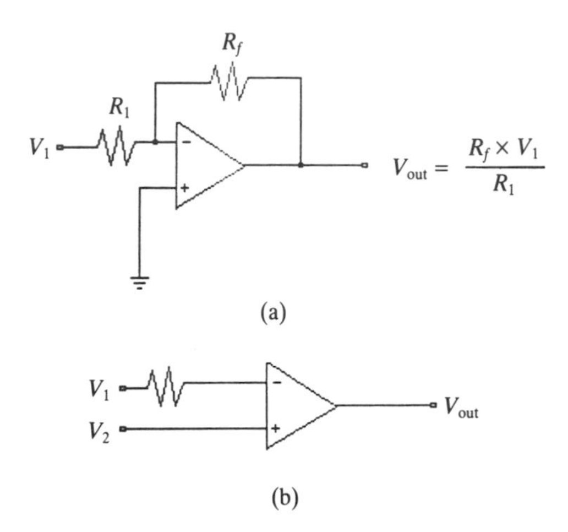

Figure 5.1 (a) Closed-loop and (b) open-loop configuration.

5.2 Feedback Configurations

An op amp is tested in a closed-loop configuration. Some form of feedback is required to set the gain and control all of the device functions. The feedback components also determine the frequency response of the device. In fact, almost all applications of an op amp demand a set of feedback configurations that can be either voltage or current and positive or negative. As a result of the already present DC offset in any type of op amp, the smallest applied voltage can cause the output of the device to go to saturation and overload the device if there is no feedback mechanism (this phenomenon is called the *railing condition*).

The *closed-loop gain amplitude*, which is directly related to the passive components R_1 and R_f in Figure 5.1, is set as a consequence of the feedback components. Several different feedback connection configurations (creating different output response) can be formed using either of the amplifiers represented in Figure 5.1.

5.3 Basic Op Amp Internal Structure

Figure 5.2 depicts the internal structure of a basic, emitter-coupled, differential op amp. In the fabrication of this type of op amp only resistors and transistors that are easily manufacturable on a chip are needed. These two types of elements are frequently used in linear integrated circuits.

The two halves (A and B) of the circuit in the basic op amp structure depicted in Figure 5.2 must be identical and symmetric: the elements in each half must be exactly the same. Transistors Q1 and Q2 must be exactly alike and both must

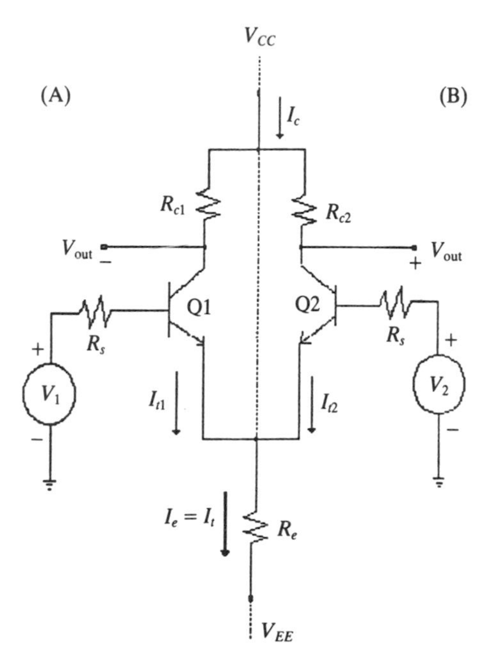

Figure 5.2 Basic internal structure of an op amp.

have their emitter connected to a common resistor R_e. Because of this common resistor, the circuit output is affected by either or both inputs.

One of the greatest advantages of integrated op amp circuits is that the effect of temperature on transistors Q1 and Q2 is almost the same, since they are implanted so close together on the die.

Two input signals (V_1, V_2) are applied, one at the base of each transistor, and the output (the voltage between the collectors of Q1 and Q2) is either V_{out+} or V_{out-}. The two collectors may be connected to other transistors in some applications, but the collector connection should always prevent Q1 and Q2 from saturating.

The current flowing through resistor R_e (sometimes known as *tail*) is shown as I_t or tail current; R_e and V_{EE} form a constant-current source. The amount of tail current is given by

$$I_t = \frac{V_{EE}}{R_e}$$

In the case where Q1 and Q2 are identical, the current is

$$I_{t1} = I_{t2} \qquad \text{and} \qquad I_t = I_{t1} + I_{t2}$$

Approximately,

The supply (V_{CC}) minus the voltage drop across R_{c1} equals the DC voltage from the collector of Q1 to the ground (V_{c1}). The supply (V_{CC}) minus the voltage drop across R_{c2} equals the DC voltage from the collector of Q2 to ground (V_{c2}).

These relations are given by the following formulas:

$$V_{c1} = V_{CC} - I_c R_{c1}$$

$$V_{c2} = V_{CC} - I_c R_{c2}$$

An op amp is operating in *common mode* if the same input signal is applied to both inputs; Q1 is identical to Q2. This resullts in an output voltage of zero.

$$R_{c1} = R_{c2} \qquad \text{then} \qquad V_{c1} = V_{c2}$$

When the supply voltage V_1 is equal to V_2, the output voltage must theoretically be zero. Any voltage present at the output is called the *offset voltage.* Offset voltage is discussed in Secction 5.10 on the VOS test.

An output voltage should occur when V_1 is not equal to V_2, or when the V_1 supply is more positive than the V_2 supply. The latter condition causes more current to flow through the collector of the Q1 transistor.

If the transistors are exactly identical, an increase in collector current will not cause the output voltage to vary with temperature. However, there is always some output drift because it is virtually impossible to exactly match sides A and B of Figure 5.2. Drift is best decreased at the design level by curbing the variation of the collector current with temperature.

5.3.1 Analysis of Basic Op Amp

Consider Figure 5.3. Two major factors are considered in AC operation of an operational amplifier, *loop gain* and *slew rate.* The loop gain should cover the highest frequency for the intended use of the device, and the slew rate should cover a range of frequencies without distortion. The high-frequency op amp produces high output current that can drive a capacitive load or any side-effect of transmission line capacitance at the device output. Slew rate is discussed in Sections 5.8 and 5.8.1.

If all of the DC sources are assumed to be zero, V_{CC} is grounded and the I_T current source is opening. Two AC sources (V_1 and V_2) now drive the circuit and the analysis for the output signal can be done based on the supply condition.

5.3.2 Supply V_1 is On Only (Noninverting Amplifier)

When a positive signal is applied to the inverting input, a signal that is negative with respect to ground will appear at the output terminal of the device.

In Figure 5.4, a noninverted sinusoid appears at the collector of Q2, which acts as a common-emitter amplifier. AC voltage appears between the collector of Q1 and ground. Transistor Q2 behaves as a common-base amplifier because it receives the emitter current of the Q1 transistor. As a result, a signal that is in

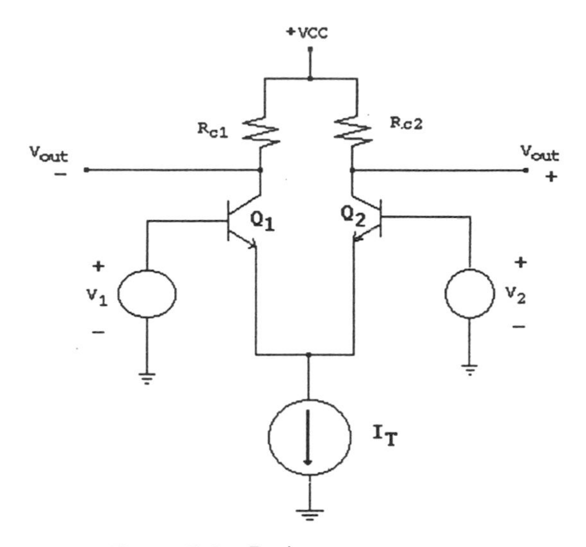

Figure 5.3 Basic op amp structure.

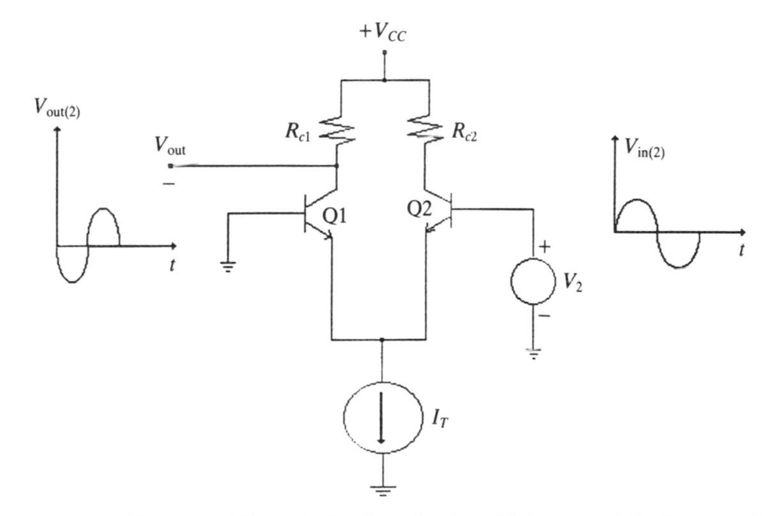

Figure 5.4 Op amp AC analysis Case 1 when V_1 is on and V_2 is grounded.

phase with the applied input signal appears at the collector of Q2. The AC output voltage signal is measured between the collectors. The input and output are said to be in phase with each other.

In a grounded-emitter amplifier the phase shift is 180°. Algebraically the output signal will have the same amplitude as the input signal. Therefore, when only V_1 is the active source,

$$V_{O1} = A V_{Q1}$$

where A is the unity gain in our example.

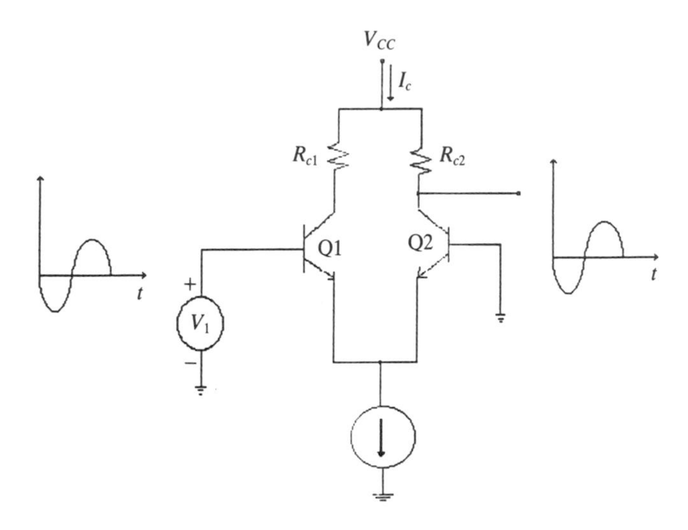

Figure 5.5 Op amp AC analysis Case 2, when V_2 is on and V_1 is off.

5.3.3 Supply V_2 is On Only (Inverting Amplifier)

In Figure 5.5, the voltage measured between the base and the emitter of Q2 would be a little smaller in magnitude with respect to V_1 supply. Q2 acts as a common-emitter and Q1 acts as a common-base amplifier. The Q1 collector displays an inverted signal.

The AC output is the algebraic difference between the collector voltages:

$$V_{O2} = -AV_{Q2}$$

On the basis of the superposition theorem, the output voltage is the sum of the input sources when both input supplies act simultaneously, or

$$V_{\text{out}} = V_{O1} + V_{O2}$$

$$V_{\text{out}} = AV_{Q1} - AV_{Q2}$$

A differential op amp's output voltage can be the source for another stage. A positive V_1 acting alone produces a positive output voltage, which makes it a noninverting input. The opposite is true of a positive V_2 acting alone; it produces a negative output (inverting input).

5.4 Common Mode Signal

The two halves of the differential amplifier shown in Figure 5.6 are identical. The input signal is connected to both inputs. The collector voltages of both transistors Q1 and Q2 stay the same as the input signal rises toward positive. Both transistors receive the same power and turn on with similar driving force. This signal is called *common mode signal* and the AC output voltage equals zero.

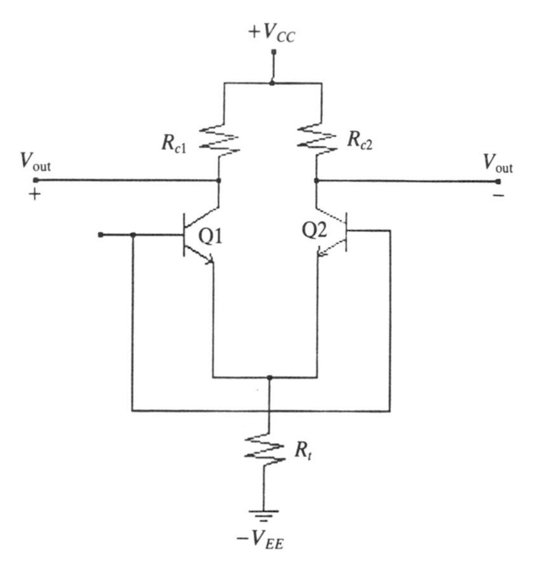

Figure 5.6 Common-mode input configuration.

This common mode input signal is a form of op amp testing used to verify the proper balance of each half of the op amp internal circuit.

The relationship between gain and the common input/output voltage of an op amp in terms of common mode rejection ratio (CMRR) is

$$\mathrm{CMRR} = A\frac{V_{\mathrm{in}}}{V_{\mathrm{out}}} = \frac{A_d}{A_c}$$

where V_{in} and V_{out} are the common input and output voltages, and

A_d = differential input voltage gain

A_c = common mode signal gain

The CMRR shows that the common mode voltage has been rejected or that there is asymmetry in the op amp internal circuit. In Figure 5.6, for example, if V_{in} is 2 V we should get 0 V at the output; a nonzero output level indicates a degree of asymmetry.

Suppose A is 100 and $V_{\mathrm{out}} = 0.02$ V, then the voltage ratio is

$$\frac{100(2\ \mathrm{V})}{0.02\ \mathrm{V}} = 10\,000$$

CMRR is customarily expressed in dB. The above value in dB is

$$\mathrm{CMRR} = 20\log(10\,000) = 80\ \mathrm{dB}$$

This indicates that the output common mode signal will be 80 dB below the output differential signal.

CMRR is a measure of the quality of an op amp; the larger the value of CMRR, the better the op amp. Ideally CMRR is defined to be infinite.

5.5 Input Impedance

An ideal amplifier should have an infinite value of Z_{in} between its input terminals. It actually has a very large value of Z_{in}. The influence of this input impedance should be reduced, because it may be nonlinear and could affect the op amp's linear response. In theory, the output of any linear circuit can be Thévenized; i.e., the output of an op amp can be considered as the Thévenin voltage source (V_{th}) of

$$V_{th} = A_{v1} - A_{v2}$$

Input impedance (Z_{in}), output impedance (Z_{out}), and also the gain of an op amp are shown in Figure 5.7. These are a few of the most important parameters in any op amp that are not usually shown on the conventional device symbol. These parameters explain the practical behavior of this group of electronic devices when their outputs are the driving forces for later stages. The Z_{out} and $A_{v1} - A_{v2}$ voltage source represent a Thévenin circuit at the rear end of the symbol.

An op amp's input impedance is not tested in semiconductors. The input bias current is usually measured instead because of the close relation between the input impedance and the input bias.

5.6 Single-ended Output

A single-ended output is measured between either output and the ground. With this form, the voltage gain is decreased by half, because it uses only half the available output voltage, and the configuration is sometimes used in the later stages of an amplifier. The advantage of single-ended output is that it reduces drift at the output. Figure 5.8 illustrates single-ended output. In this configuration,

$$V_{c2} = A\frac{V_{in}}{2}$$

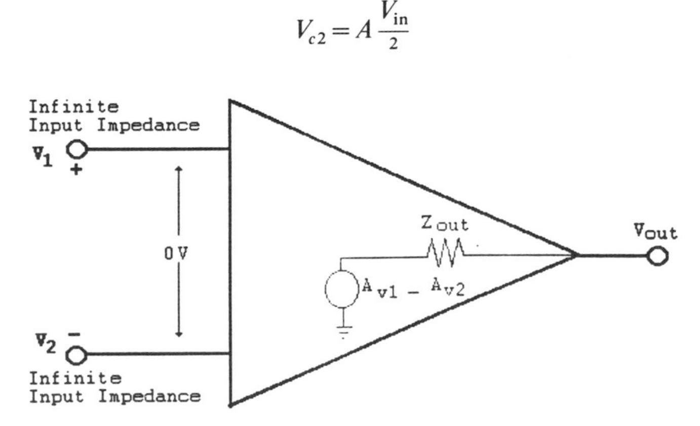

Figure 5.7 Input impedance and output circuit.

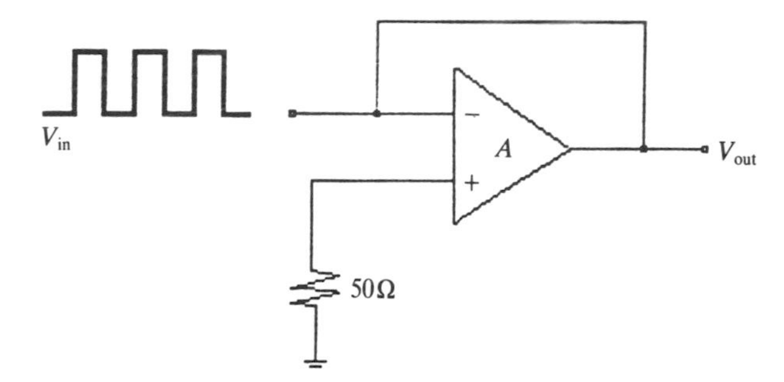

Figure 5.8 Single-ended output schematic.

5.7 Input Bias Current (I_B)

Although the input of an ideal op amp draws no current, there will be a minuscule current at the base of Q1 or Q2 if the op amp is composed of bipolar transistors. This temperature-dependent current is called the *input bias current* (I_B), which can bring about major problems in high source input impedance, but can be neglected at low source. This current can be quantified by forcing it through a resistor and measuring the voltage across the resistor. The voltage created by I_B across the resistor is in series with VOS; therefore, the measured voltage will be

$$V_{\text{total}} = \text{VOS} + I_B R$$

If the value of VOS is known, the I_B value can be determined, but since this current is very low, the value of R must be precise. The input bias current shown on the data sheet is an average of the two input currents. It is an approximation of each input current. A general guide is that the smaller the input bias current, the greater the possibility of balance between the two sides of Figure 5.2.

5.8 Slew Rate (SR)

Slew rate in volts per microsecond can be defined as the highest rate at which the output signal voltage changes with time in response to a step signal input. In other words, slew rate is a parameter that shows how fast the output voltage changes with time. The output voltage change is caused by the internal capacitor that is used to adjust the small-signal frequency response of the device.

Ideally, an op amp should respond to all frequencies uniformly. For example, if the output of an op amp is made to vary from one extreme to another by applying a step signal at the input, the output should go through a smooth transition between the two extremes; i.e., the slew rate is the speed at which the output signal changes. Slew rate is defined as

$$\text{SR} = \frac{\text{maximum charging current}}{\text{capacitance}}$$

5.8.1 Sources of Slew Rate

Slew rate is the most important specification affecting the dynamic operation of an op amp and poses a severe limitation on the large-signal operation of the device. The current limiting and saturation characteristics of the op amp's internal structure produce the slew rate. If the output signal is smooth and noise-free, the highest maximum frequency at which the device can operate for any given voltage swing is determined by the slew rate.

The low-frequency response of the circuit to drive either the internal or external capacitative load is adjusted using a capacitor in the device's internal circuitry. At high frequencies the current available to charge or discharge this capacitor is expended and the output signal becomes distorted. The speed at which this capacitor charges will decrease when the charging current is too small to respond quickly.

The slew rate is most severe in noninverting mode at unity gain. Although the network configuration used to measure this parameter is dependent on the type of application, the measurement condition described below is based on the unity condition case since it is the worst case affecting this parameter.

5.8.2 Slew Rate Test

Procedure

Set the signal generator to produce $\pm 10\,\text{V}$ sine or $10\,\text{V}$ square wave with sharp edges. The frequency of this pulse train should be slow enough to provide sufficient time for the output to slew from one limit to the other. The amplitude of this pulse must be lower than the maximum operating input voltage.

1. Connect the $\pm 10\,\text{V}$ pulse to the input.
2. Measure the output voltage.

$$\text{SR} = \frac{\Delta V}{\Delta T}$$

T is actually the rise time that is measured from 0 to 100 percent of the $\pm 10\,\text{V}$ output.

Figure 5.9 shows a graphical representation of this test. When a large signal is used to measure the slew rate and noise is the interfering factor, the configuration in Figure 5.10 is useful.

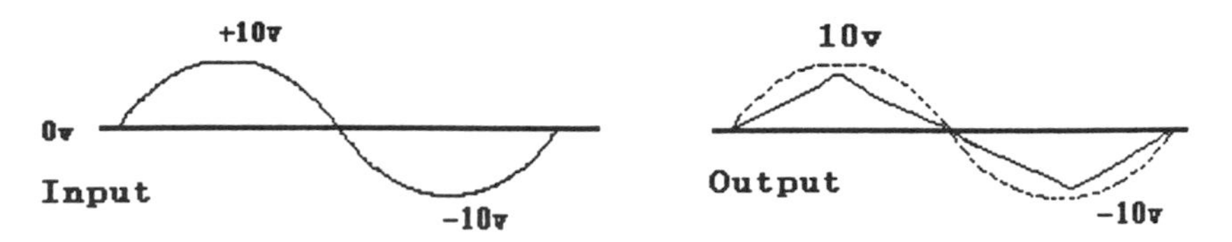

Figure 5.9 Slew rate distortion of sine wave.

The procedure for measuring slew rate in this form is the same as outlined above. The schematic in Figure 5.11 can be used to measure both the rising and the falling slew rate.

$$SR_{\pm} = \frac{\Delta V_{\pm}}{\Delta T}$$

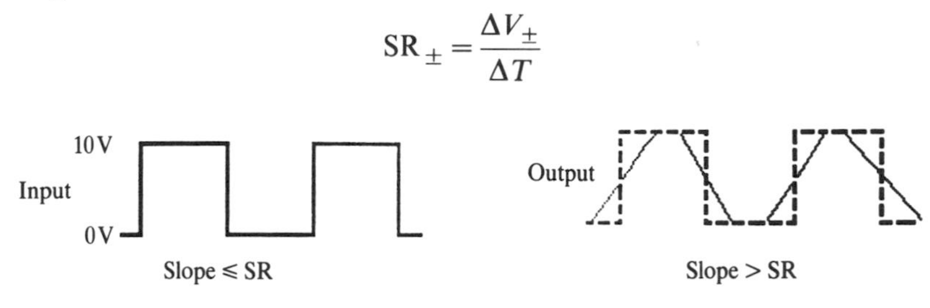

Figure 5.10 Large signal overcomes noise in SR testing.

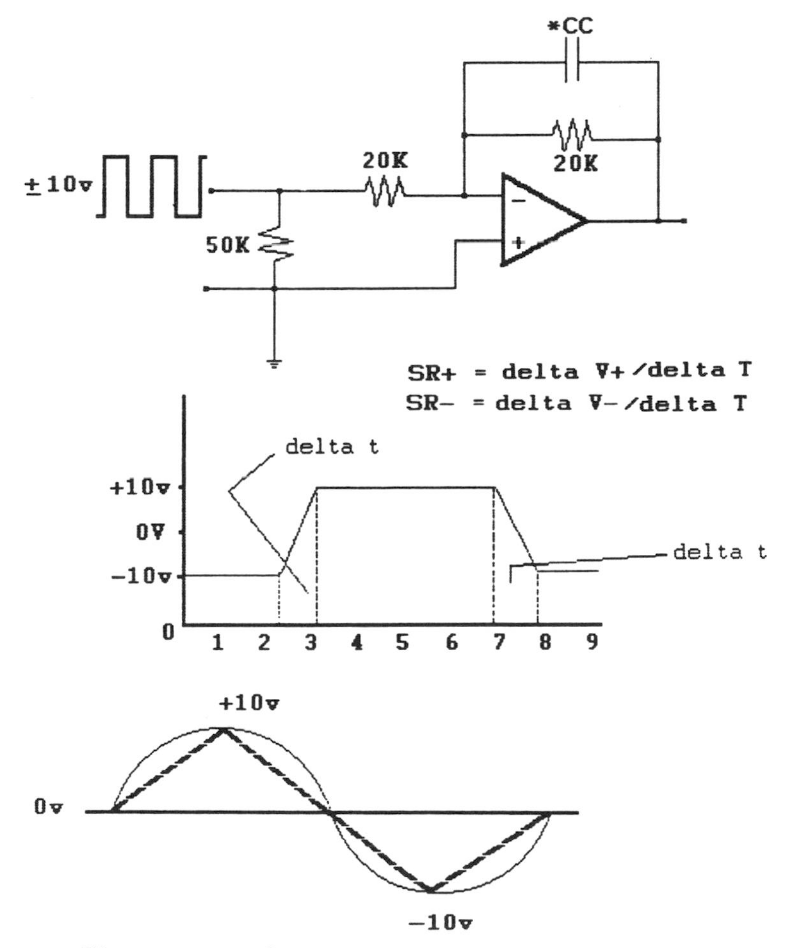

Figure 5.11 Input and output response in slew rate test. *CC is a small (nF range) capacitor to suppress possible DC noise.

5.9 Op Amp DC Measurement (Nulling Circuit)

Similarities among various types of op amps have enabled test engineers to develop a universal test circuit to perform all the device-standard DC measurements. Universal test circuits may look different in appearance or the values of the components, but they work essentially the same.

The value of input or feedback resistors and other components that may be required to be added to the circuit are dependent on the accuracy, limit size, and, most of all, the characteristics of the chosen nulling op amp.

5.9.1 Practical Nulling Circuit

All relays used in this circuit are normally open. Figure 5.12 shows a typical working circuit that was implemented to measure very low offset voltage and current. R_1 is the resistive load.

The output of the DUT is connected to the noninverting input of the nulling op amp (N) and the inverting input of the nulling op amp is set to control the loop.

The nulling op amp is fed negatively to maintain equal input voltage.

The output of nulling controls the DUT input through the feedback network and keeps the output of the DUT to the voltage applied at terminal D. Nodes A, B, C, D, and E are connected to the force and measurement units of the ATE.

The nulling circuit should be adjusted to cooperate with the DUT. This requires adequate knowledge of the DUT. For example, the frequency response of the DUT must be known before any step is taken to stabilize the AC response of the circuit. The output voltage of the nulling op amp after the application of biasing voltage and before the DUT testing must be set to 0.0 V. The technique for obtaining this very important basis is dependent on the type of op amp chosen. The manufacturer's data book is the best guideline. Accurate measurement will be impossible in the presence of any VOS due to these adjustments.

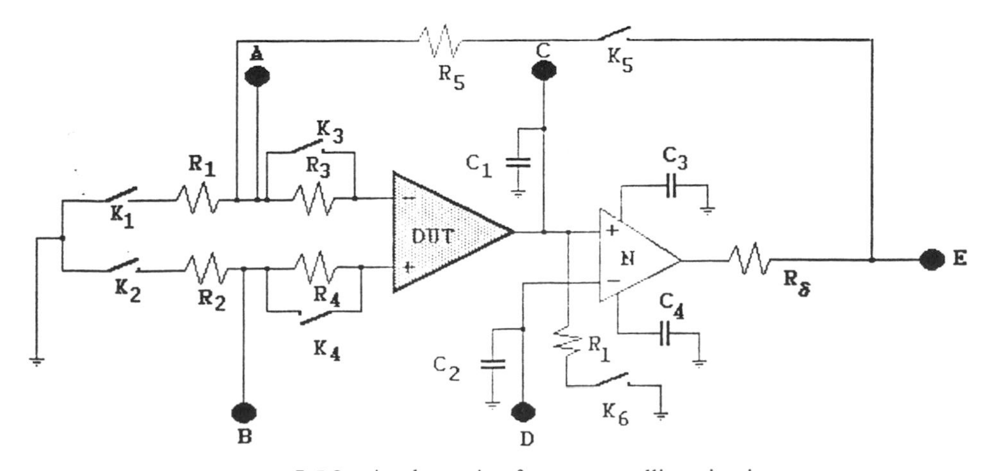

Figure 5.12 A schematic of op amp nulling circuit.

The technique for cleaning noise from the output is standard and based on the appropriate combination of *RC* circuits required to filter noise. The power supply compensation, especially the nulling operating voltage, is of special concern. Any small glitch or ringing on that supply is magnified at the output. The nulling op amp has its own characteristic that demands its own compensation components, which should ultimately be added to the circuit. The recommended compensation elements are described in the device data book.

In any case, the nulling op amp N must, in many ways, be more accurate and have better frequency response than the DUT. Having the above knowledge of the nulling circuit along with the basic op amp operation, we now proceed with all of the op amp DC parameters.

5.10 VOS Test (Input Offset Voltage)

VOS is the differential input voltage needed to drive the output to 0 V and is a temperature-dependent parameter.

5.10.1 Sources of VOS

In an ideal op amp, the output voltage should go exactly to 0 V when the differential input voltage is set to 0 V. Real op amps do not show such an idealized scheme. The main causes of VOS in a real op amp are as follows.

1. The op amp internal circuit imbalances that are a result of the manufacturing process. These imbalances cannot be totally avoided (refer to Section 5.3).
2. The offset of DC input current to the base of the bipolar transistor (for TTL type) or the gate-leakage current (in MOS type).

VOS is an important factor in high-gain circuits. Some small voltage should be applied to the input that is the result of the op amp DC noise sources and cannot usually be detected at the DC output signal.

When one op amp is used to provide gain, these noise signals are amplified and affect the output signal in their own ways. This is not a big problem when an op amp is used in an AC application, where these noise signals move toward the Q point of the transistor. The magnitude of the voltage swing cannot be significantly degraded by this noise.

Methods to diminish the effects of VOS in DC application of an op amp include the following:

1. Use an op amp that has the lowest VOS.
2. When the DC input current seems to create this problem, use an op amp with a smaller input current capability, such as Bi-FET.

Two methods of measuring VOS are

1. Direct measurement on the DUT.
2. VOS measurement by the aid of a nulling circuit (refer to Section 5.11).

One method of VOS testing (though unpopular because it is time-consuming yet imprecise) is to apply a small varying voltage at the input, monitor the output, and hope for exactly 0 V. The input at which the zero output results is the VOS. See Section 3.14 on the voltage hysteresis (V_{HYS}) test.

5.10.2 Direct VOS Measurement

Note. In this method there will be an error of a few percent due to the likely inability to set the output at exactly 0 V before measurement. The value of the input resistor must be small to prevent measurement error due to the resulting input current.

The test configuration is given in Figure 5.13.

5.11 VOS Test Using Nulling Circuit

Figure 5.14 shows the location of relays and nodes: these are referred to in several of the tests described below.

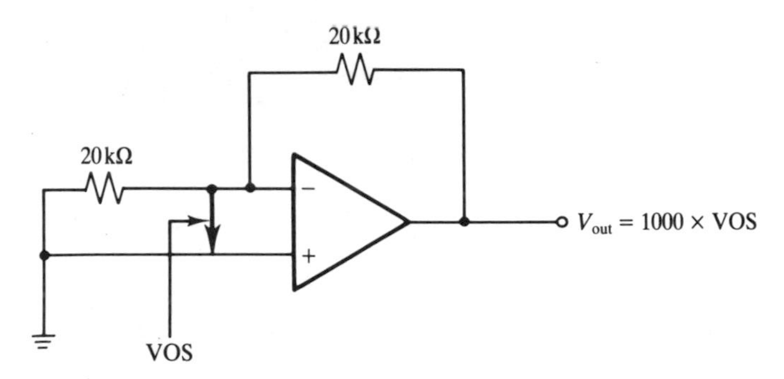

Figure 5.13 Circuit for direct VOS measurement.

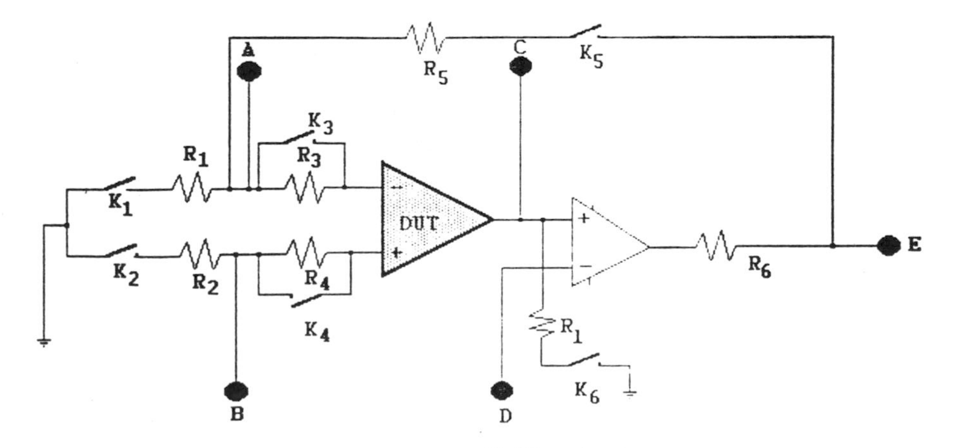

Figure 5.14 Nulling circuit.

Definition

VOS (offset voltage), is the voltage that if applied to the input of an operational amplifier will produce 0 V at the output.

No load is required for this test.

Procedure

1. Close relays $K1$ through $K5$.
2. Set nodes A, B, C and D to 0 V.
3. Measure voltage at node E.
4. Divide the measured voltage by the circuit loop gain. Test the resulting measurement against the specified limit.

Important Note. If there are a few other op amps on the same die that share the same supply voltage, care must be exercised to tie the I/O of the rest to ground. Untied I/O creates severe instability problems in DUT testing.

5.12 IOS Test (Input Offset Current)

Definition

IOS is the difference in the currents into the two input terminals of an op amp when the output is at 0 V.

Discussion

The DC input current is the current that flows through the inputs of an op amp when no voltage is applied. The sum of the two bias input currents must be zero, otherwise it causes serious unbalance when the op amp is used in a feedback configuration. The nonzero value of the differential current is called input offset current (IOS).

$$\mathrm{IOS} = \mathrm{ABS}(I_{B-} - I_{B+})$$

Figure 5.15 illustrates the concept.

In a temperature-dependent device, the IOS will vary as the internal temperature of the device changes. This temperature-related shift of IOS is called

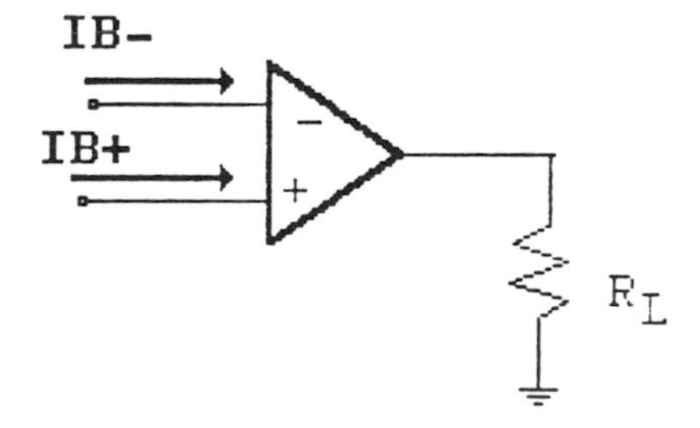

Figure 5.15 IOS diagram.

input offset current drift, which can be determined at any desired temperature. The relationship between the voltage and current drift with respect to temperature is

$$V_{dr} = \frac{\Delta \text{VOS}}{\Delta T} \quad \text{and} \quad I_{dr} = \frac{\Delta \text{IOS}}{\Delta T}$$

In practice this parameter is about 10 percent or less of the bias current (I_B) in well-matched op amp inputs. These unequal base currents create a false input signal when they flow through DC returns. The smaller the IOS, the better the op amp.

To measure the input offset current, connect a resistor in series with each output and force the bias current through each resistor. This will produce a bias voltage source in series with each input.

Procedure (Using Nulling Circuit)

1. Close relays K1, K2, K4, and K5.
2. Set nodes A, B, C and D to 0 V.
3. Measure voltage at node E (V_1).
4. Close relay K3 and open K4 (other relays are the same as in Step 1).
5. Measure voltage at node E (V_2).
6. Take the absolute value of ($V_1 - V_2$).
7. Divide the result by the total loop gain (sum of the loop resistors).
8. Convert the result into current using the R_3 and R_4 values.
9. Test the measured value.

5.13 VON Test (Negative Output Voltage Swing)

Definition

The capability of an op amp output voltage to reach its highest negative value (close to V_{CC} or V_{DD}) by the application of a small voltage (1–2 V) at the inverting input terminal. This is also called *negative railing* and is caused by the saturation of the internal transistor.

Discussion

Even though the relationship between the output voltage swing and the slew rate is not given in most manufacturers' data sheets, these two parameters are closely related and dependent on the compensation capacitors.

The application of 1 V signal at either input should give 0 V at the output (neglecting VOS); otherwise the output voltage swing is less than V_{CC} or V_{DD} by the amount of internal op amp transistor saturation, which requires about 1 V when a nonzero voltage is applied at either input.

Procedure

This test is performed with some specified load.

The following steps measure the output voltage swing by the application of the assigned voltage to the inverting input.

1. Set all relays in the nulling circuit to open.
2. Connect FMU (force measurement unit) of the ATE to node A (inverting input).
3. Connect FMU to node B (the inverting input).
4. Connect FMU to node C (op amp output).
5. Force a defined voltage to node A while node B is forced to GND.
6. Measure the output voltage at node C.

The measured voltage must be close in absolute value to $-V_{DD}$ or $-V_{CC}$.

5.14 VOP Test (Positive Output Voltage Swing)

Definition

The capability of an op amp output voltage to reach its highest positive value (close to V_{DD} or V_{CC}) by the application of a small voltage to the noninverting input terminal. This is also called *positive railing*.

Discussion

The discussion for VON is applicable to this test. The only difference is that the output voltage will swing to the positive direction.

Procedure

A specified load is involved in this test.

The same test procedure as for VON is applied for this test with the exception that node A should be forced to GND and node B should receive the specified forcing signal. The output voltage must be close to $+V_{CC}$ or $+V_{DD}$.

5.15 CMRR Test (Common Mode Rejection Ratio)

Definition

CMRR, or in-phase rejection, is the ratio of output offset voltage (VOS) to the change in common mode voltage and is normally expressed in decibels. The ability of the op amp to cancel the common mode signal is called *common mode rejection*.

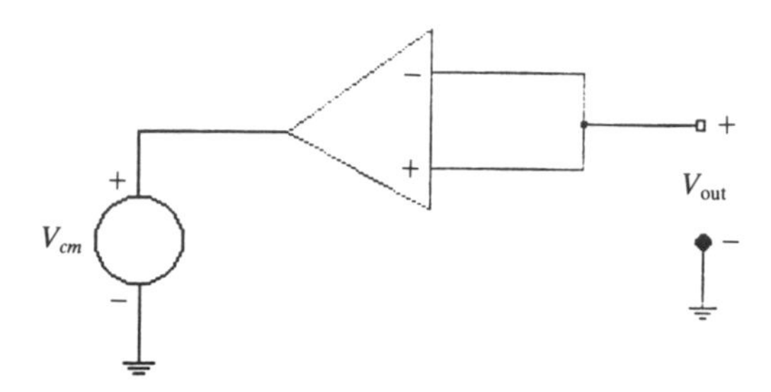

Figure 5.16 CMRR configuration.

Discussion

This parameter is inversely proportional to frequency. The source of CMRR is again the unbalance in the internal structure of the op amp. Another way of describing this parameter is to say that CMRR is the ratio of common mode noise voltage to the differential noise voltage.

When the internal circuitry of an op amp is well balanced, the absence of noise helps the elimination of CMRR. In a well-balanced op amp, the value of this parameter ranges from 60 to 80 dB.

The main factors that aggravate CMRR are load, source, signal load, and, most importantly, the impedance. Proper decoupling and grounding of the circuit helps impedance reduction. Figure 5.16 shows a simple method of CMRR configuration in an op amp.

$$\text{CMRR} = \frac{V_{\text{out}}}{V_{\text{cm}}}$$

Procedure

No load is required for this test.

There are many ways to measure this parameter. One easy method that also eliminates the need for precision resistors for circuit balance is discussed here. The idea is to keep the DUT output at the midpoint of the voltage supply. Refer to the nulling circuit of Figure 5.14.

1. Close relays K1 through K5.
2. Force 5 V to V_+ pin of the DUT (assuming the internal voltage is ± 15 V). This will apply a common mode voltage of 10 V.
3. Force -25 V to the V_- pin of DUT (assuming the initial voltage is ± 15 V). This will apply a common mode voltage of 10 V.
4. Force 2 V at node D.
5. Measure the voltage at node E (VOS1).
6. Force 15 V to V_+ node (this will apply a common mode voltage of -10 V).

7. Force -15 V to the V_{-} node (to create a common mode voltage of -10 V).
8. Set V_C to 0 V.
9. Measure voltage at node E (VOS2).
10. Convert the result into decibels.

5.16 SR Test (Slew Rate)

Definition

Slew rate, or rate timing, is the maximum rate of change of output voltage in response to a square wave or any step input signal. Slew rate is expressed in volts per microsecond.

Discussion

Slew rate (SR) is caused by the capacitor used to adjust the small-signal frequency response of the device in the internal op amp structure. In other words, SR is a parameter that shows how fast the output voltage changes. Refer to Section 5.8.1 for more detail.

Procedure

1. Set the signal generator to generate ± 10 V sine or 10 V square wave with sharp edges. The frequency of the pulse train must be low enough to provide sufficient time for the capacitor charge/discharge. The amplitude of the pulse must also be lower than the maximum device operating voltage.
2. Force the pulse train to the device input.
3. Measure the output voltage.
4. $SR = \Delta V / \Delta t$. Time t is the rise time measured from 0 to 100 percent (according to definition it is from 10 to 90 percent of rising or falling edge) of the ± 10 V output.

5.17 AV (Gain) Test

Definition

An increase in the output signal is called gain and is expressed as the ratio of changes in the input and output voltages. The unit of gain measurement is the decibel (dB).

Procedure

The gain test is always specified with some resistor load.

1. Set K1 through K6 to closed. K6 connects the desired load to the DUT output.
2. Connect the ATE measurement unit to node A.
3. Apply -5 V (can be any desired value) to node D.

4. Measure voltage at node C (V_1).
5. Apply +5 V (can be any chosen voltage) at node D.
6. Measure voltage at node C (V_2).
7. Calculate the absolute value of $V_1 - V_2$.
8. Divide the result by loop gain and multiply by 10 (the voltage range, in our case −5 V to +5 V).
9. Convert the result into decibels.

Note. A good method for checking the linearity of an op amp transfer function is to check the gain for two different swings separately. For instance, instead of measuring gain from −5 to +5 V, in our case we can measure the gain from 0 V to −5 V and then from 0 V to +5 V and compare the results.

5.18 PSRR Test (Power Supply Rejection Ratio)

Definition

This verifies the insensitivity of the output to changes in the power supply voltage (V_{CC} or V_{DD}). In other words, it is the ratio of change of VOS to the change in power supply voltage.

Discussion

The device should be able to accommodate fluctuations in power supply voltage without change in the output. If the output of one stage of the amplifier draws a variable amount of current, then the supply voltage can be affected and the change may affect other amplifiers that are sharing the same supply.

This effect is another form of crosstalk that creates current instability. A method of preventing this problem is to decouple the power supply. PSRR is typically about 20 μV for every volt change in power supply. The measurement unit for PSRR is μV/V or decibels.

Procedure

A defined load is required for this test.

1. Close relays K1 through K5.
2. Set nodes A and B to 0 V.
3. Set the power supply voltage to the defined voltage.
4. Measure voltage at node E (V_1).
5. Set power supply voltage to a value different from that which has already been defined.
6. Measure the voltage at node E again (V_2).
7. Take the absolute value of ($V_1 - V_2$).
8. Divide the result of Step 7 by the loop gain and convert to decibels.
9. Test the measured value.

5.19 Noise in Op-amp Circuits (Crosstalk)

The supply voltage in a digital or analog circuit can cause a variety of problems that result in circuit malfunction. One of the adverse effects of current spikes in a power supply occurs when these spikes, due to the limited source resistance, become voltage spikes.

The voltage spikes may falsely trigger a flip-flop or any other logic gate in a digital system. This effect is called *crosstalk* and is simply the effect of an unfiltered power supply on the operation of the circuit. In the case of op amps, when the output of one stage draws a different amount of current due to the power supply change, malfunction is caused in other op amps that are sharing the same supply. Bypass capacitors on different parts of the power rail can lower the magnitude of the supply voltage spikes by reducing line impedance.

Likewise, interference can be caused between different channels by other undesired frequencies on the line due to a magnetic or electrical field induced between conductors. The current change in a wire conducting a signal can induce current in an adjacent conductor. The induced current is proportional to the rate at which current changes and also to the length of the conductor.

Unlike other noise sources, crosstalk is expressed in dB and the reference point is 90 dB. For instance, if system X is being induced by system Y at 70 dB lower power, then the cross-talk from system X to system Y is

$$90 - 70 = 20\,\text{dBX}$$

5.20 Oscillation

Instability problems in circuits containing amplifiers are very common, even though today's amplifiers are so well designed that the inherent device components hardly create a problem. Components that are externally attached to the device are the main causes of oscillations.

There are many complex mathematical relations for verifying oscillation and finding an appropriate theoretical compensation, including Nyquist plot, root locus, and Bode plots. None of the methods is quick or easy. Test engineers need a quick and easy way of taking care of such problems—the test floor is not the right place for mathematical calculations. Since the feedback circuit dominates all the behavior of an op amp and controls the gain by the defined values, it is one of the main factors that must be focused on when instability arises.

Feedback can be from either negative or positive input or from a combination of the two. The real problem occurs when the phase shift feedback at the positive input becomes negative or vice versa. Many factors are involved in the occurrence of this phenomenon. The following describes and discusses each, with information deemed useful by the author. In this section the most important causes of oscillation are given along with easy methods for finding the exact cause and then the appropriate action to suppress them.

5.20.1 Power Source

Ideally the AC impedance between the power supply and ground should be zero for a guaranteed noiseless design. In reality this is not practical because the power supply itself, as well as wires connecting it to the op amp circuit, have inherent resistance and inductance.

This problem can be solved by connecting a ceramic or tantalum type capacitor between the power supply and the ground of the op amp and very close to it, especially in high-frequency applications. This capacitor will short the circuit in the high-frequency range at which the op amp may be capable of creating gain. In the absence of this short circuit, the AC voltage gain created at the power supply can be fed back to the amplifier and create oscillation. This additional bypass, if it is large, will reduce transient ground noise and create a longer recovery.

With respect to capacitive load, any such load as transmission or coaxial cable being fed by an emitter follower can be the cause of oscillation at high frequencies when the power supply is not properly decoupled. The cause of oscillation is phase lag. The solution to this type of problem is to decouple the supply source with a capacitor or an *RC* filter connected directly from the supply to the output pin and the ground. The value of the resistor in this filter is small, and is dependent on the frequency of oscillation.

5.20.2 Manufacturers' Recommendations

Keep a close match with compensations that are given by the op amp manufacturer in the device data book. The frequency response compensation given in the data book is the most suitable for any recommended application. In low-frequency or DC applications, it is best to overcompensate the circuit for better stability.

5.20.3 Extra Inverting Input Resistance

In mathematical terms, excessive resistance between the ground and the negative input creates a pole in the loop gain. The problem increases greatly when the input resistance of the device is low. Use of a capacitor to connect the noninverting input to the ground can solve this problem. The correct value for this capacitor can be found quickly by trial and error during the test debugging process; its value depends on the input signal frequency and on the combination of input and feedback resistors.

5.20.4 Phase Shift by Capacitive Load

Phase lag and oscillation (sometimes called ringing) are caused by excessive load capacitance. A lagging circuit is created if a capacitor is connected between the output and the ground, placing the ground in series with the feedback. The output resistance of any op amp depends on the input signal frequency; the resistance reaches its largest value when the op amp is set at unity gain. The effect of load capacitance is aggravated when the scope probe is attached to the device output pin; this connection creates a pole in the loop gain and diminishes the circuit bandwidth. Certain designs require some capacitance at the output. The manufacturer's data book is the best reference for determining this.

5.20.5 Phase Control Feedback

Stray capacitances at the op amp input cause phase lead and thereby phase lag in the feedback network. Each op amp input to the ground and between the capacitors (C_1, C_2 and C_3 in the Figure 5.17) produces stray capacitances. The feedback factor of such a circuit will have two poles and a zero. One solution to phase lag is to connect a small-value capacitor (in the picofarad range) in parallel with the feedback resistor to create a zero in the feedback gain. Figure 5.17 shows this compensation capacitor.

5.20.6 Special Op Amp Types Requiring Appropriate Capacitors

In some special types of op amps with accessible balance terminals, negative feedback will become positive if capacitance occurs between the balance and the output terminals. To prevent this, balance terminals should be connected together when not in use and a nanofarad-range capacitor should be connected between each when they are being used.

5.21 Noise in Different Types of Operational Amplifiers

A high-gain amplifier's output will have some level of noise even before any of the input signals are amplified. The evidence of noise levels can be seen on an

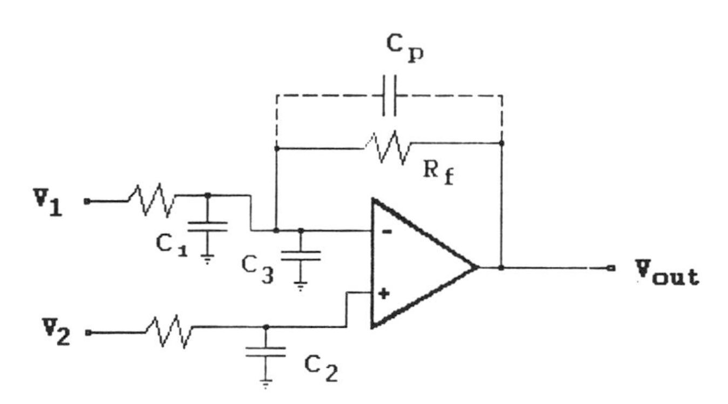

Figure 5.17 Capacitors C_1, C_2 and C_3 cause phase lags C_p cancels them.

oscilloscope as fuzzy traces around some DC levels. Depending on the kind of input structure it has, any type of op amp generates its own characteristic noise. The following are the most important sources of inherent noise, listed according to op amp structure (these are design problems and of no concern to test engineers).

5.21.1 Bipolar Type Op Amp

The sources of noise in this type of op amp are base-spreading resistance and base current of the internal transistors (see also Section 4.15). The bipolar structure is used when the source resistance does not exceed 100 kΩ. Its greatest advantages are low DC offset and insensitivity to temperature, and it is usually used in noninverting configurations.

5.21.2 FET Type Op Amp

The predominant type of noise in an FET type op amp varies according to the source resistance: thermal noise if it is low and shot noise if it is high. The noise arises from internal channel resistance and the gate leakage current.

The FET type structure is used when the source resistance exceeds 100 kΩ. In very high-impedance applications in the inverting mode, the low noise of the FET type of input structure makes it an excellent choice. However, the DC offset is not as low as that of the bipolar structure.

5.21.3 Reducing Op Amp Noise Using a Voltage Divider

A series resistor and a voltage source are connected to the noninverting input of a low-noise amplifier as in Figure 5.18.

Although any amplifier has a very high input impedance in a nonfeedback configuration, the resistance will be much higher if the op amp is used in a noninverting feedback mode.

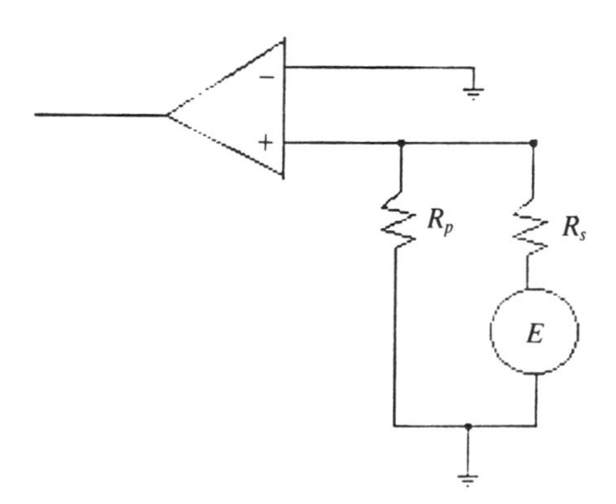

Figure 5.18 Noise reduction by voltage divider.

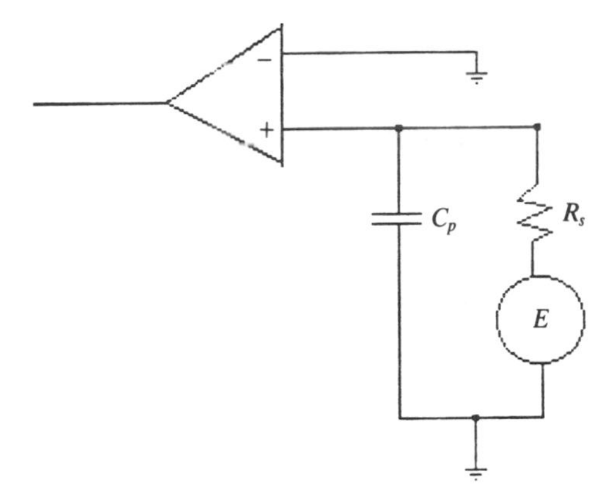

Figure 5.19 Noise-reducing C_p in an op amp.

The influence of total noise at the amplifier input can be eliminated by adding a resistor in parallel to the existing one. This resistor acts as a voltage divider and changes the input resistance. Trial and error for finding an acceptable resistor value will save considerable time: sometimes a component whose value has been determined through series of mathematical calculations does not produce the best function. This is mostly due to environmental effects on other components in the network as well as the conformation of the components.

5.21.4 Reducing Op Amp Noise Using Input Capacitance

If R_p in Figure 5.18 were replaced by a capacitor C_p, the noise voltage at the real op amp input would not change. C_p increases the ratio of input noise when frequency is increasing. Figure 5.19 shows the respective schematic.

5.21.5 Reducing DC Offset and Op Amp Drift

This method is used to reduce $1/f$ type noise, which is typically a low-frequency noise. It uses a bandpass filter with the bandpass greater than twice the low input frequency.

Another method is to use a transformer only when a low-impedance source is used. The transformer is used to boost source impedance (impedance match). This method is a less than an ideal alternative in most low-signal applications, particularly in semiconductors, because of the following inherent characteristics of most transformers:

- Core loss
- Wire resistance
- Inductance at low frequency
- Increased capacitance at high frequencies

5.22 The Decibel

The logarithmic scale used in power ratio measurement is called the decibel scale. The classical definition of the decibel is

$$\text{Value in dB} = 20\log_{10}\frac{P_1}{P_2} \tag{1}$$

When this relation is applied to voltages, it is expressed in terms of those voltages:

$$\text{Value in dB} = 20\log_{10}\frac{V_1}{V_2} \tag{2}$$

The above relation is common in amplifier and feedback discussions. When V_1 and V_2 are used in Equation (2), the common impedance and voltage levels are not needed. The result of Equation (2) will then be 20 times the logarithmic voltage ratio. For example, to find the gain (in dB) of an op amp whose value is given as 10:

$$AV(\text{dB}) = 20\log_{10} 10 = 20(1) = 20\,\text{dB}$$

A summary of some numerical values of gain and their equivalent decibel values is given in the following table.

Ratio	*dB*
0.001	−60
0.1	−20
0.301	−10
0.5	−6
0.707	−3
0.9	−1
1.0	0
1.1	1
1.414	3
2	6
3	10
10	20
30	30
100	40
1000	60
10 000	80

References

1. Robert Peas, *Troubleshooting Analog Circuits Part 8*, National Semiconductor Corp, 1991.
2 Jiri Dostal, *Operational Amplifiers*, Elsevier Scientific, 1981.
3. *Linear Op-Amp Data Book*, Fairchild Semiconductor Co., 1979.

4. Thomas M. Fredriksen, *Intuitive IC Op-Amp*, National Semiconductor Corp., 1986.
5. *Signetic Analog Manual*, Application Specifications, Signetic Corporation, 1976.
6. Charles A. Holt, *Electronic Circuits—Digital and Analog*, Wiley, 1978.
7. Jacob Millman and Christos C. Halkias, *Integrated Electronics: Analog and Digital Circuits and Systems*, McGraw-Hill, 1972.
8. John D. Lenk, *Handbook of Practical Electronics Circuits*, Prentice-Hall, 1975.
9. Jerome E. Oleksy, *Practical Solid State Circuit Design*, 2nd ed., Howard W. Sams and Co., 1981.
10. Don Lewis, *Testing Operational Amplifiers*, Fairchild Camera and Instrument Co., 1979.
11. Stan Gibilisco and Neil Sclater, *Encyclopedia of Electronics*, 2nd ed., TAB Professional and Reference Books, 1985.
12. Albert Paul Malvino, *Electronic Principles*, 3rd ed., McGraw-Hill, 1984.

CHAPTER 6

Data Acquisition Devices

Introduction

In this chapter data acquisition devices—analog-to-digital (A/D) and digital-to-analog (D/A) converters—are explained from a test engineer's point of view. Data acquisition devices that are extensively used in communication networks contain circuitry that deals with analog and digital quantities at the same time. Chapter 3 was completely devoted to digital networks and procedures for testing each parameter. In chapter 5 we chose an operational amplifier, which is a purely analog device. We will now begin to become familiar with the third class of network, which requires the combination of the knowledge acquired from those chapters. Testing the data acquisition family of devices is considered a challenge. All testable parameters in these types of devices are listed in order. The test procedure for each parameter is given in sufficient detail for a test programmer's or technician's use.

Topics Brief description of the structure of a converter to identify the internal elements and their function.

6.1 Digital signal processing and its applications in data communications
6.2. Detailed descriptions of the terms and parameters that are to be tested
6.3 Operation of a 4-bit converter, with a schematic in terms of relays, a resistor, and a comparator. This is the most effective way to give the complete information about a converter to a new electronic engineer
6.4. Data conversion process, including methods of sampling and recovery
6.5. A/D and D/A test description, pin identification, and data sheets from National Semiconductor (one of the leading IC manufacturers), which include all the parametric tests for the types of converters made by most converter manufacturers.
6.6.–6.30 Procedures for mandatory tests on data acquisition devices

An analog-to-digital converter (ADC) is a system that converts analog input signal (in voltage) to some predefined digital output sequence. The input signal can be defined in relation to the output sequence (in voltage level terms), or vice versa for a digital-to-analog converter.

For an ideal case the following relation holds:

$$D_{sig} = \frac{A_{sig}}{V_{ref}}$$

where

A_{sig} = analog signal
D_{sig} = digital signal

Some ADCs use a successive approximation register (SAR) and others use sigma-delta or half-flash techniques for converting signals.

The architecture of an ADC contains the following components:

- DAC
- Comparator
- SAR
- Clock
- Scaling register
- Voltage reference

Technologies such as MOS, bipolar and BiCMOS are used in the fabrication of these devices. The selection of any data converter must be based on the most important function one expects the device to perform satisfactorily. These main factors are accuracy, speed of conversion, and resolution. Among the secondary factors in choosing a converter are the output data format, temperature stability, required clock rating, power supply requirement, and output coding. The above factors are among those that we will focus on later in this chapter (Section 6.6 on), where we concentrate on ADC parameter measurements.

Among the dynamic functions of an ADC, the most important are the signal-to-noise ratio and the harmonic distortion, described in Chapter 7.

Understanding analog-to-digital (A/D) or digital-to-analog (D/A) conversion is essential in selecting the one that can serve the purpose best. For example, a 12-bit digital-to-analog converter (DAC) may not be accurate enough for very high-resolution use. Differential linearity, linearity, scale, gain, offset, monotonicity, and hysteresis errors should be concentrated on when choosing either of these devices.

The purpose of this chapter is to concentrate not on design detail but on the elements that are necessary for a test engineer. A test engineer must understand the operation of the device before any attempt is made to write software test code. This does not, however, mean that the design of the device must be fully understood.

Sometimes new products fail some particular test repeatedly, no matter at what condition or from which lot. This can be a good indication of design failure if the test software, hardware, and setup have been checked carefully. The test engineer should then pinpoint the problem and be able to describe the module in which he thinks the problem is centered.

6.1 Field of Application

Data aquisition devices are used in numerous fashions. The exact number of their applications and the purpose are not the main interest; however, in order to make it clear why these devices are so important in their own field, the following are the main areas in which they play vital roles.

- **Current source.** If the output of a DAC is current, then the device can be used as a cheap, easy-to-use and programmable current source. The programmable current source is used extensively in automatic test equipment.
- **Voltage source.** A simple and accurate programmable voltage source or voltage or current comparator can be obtained by properly calibrating a DAC. This function of the DAC is essential in automatic test equipment.
- **Signal generator.** A clock input to a DAC will produce an output that changes its amplitude in step and linearly. Different wave shapes can be produced by inserting a ROM between a counter and a DAC with the appropriate program burned in.
- **Digital voltmeter (DVM).** An A/D device, used properly, can measure a voltage and display that reading in the form of digital readout. An encoder/decoder can convert the readout into decimal form.
- **Function generator.** A sine or cosine wave can be generated by A/D and D/A devices when the inputs of these devices are signals that vary linearly. These functions can be digitally stored in a ROM and then used as an input to the DAC or ADC. A program can vary this simulated function as desired.
- **Multiplication and division.** A DAC can be used as a multiplier by varying its reference voltage. Also, division can be carried out if this device is inserted in a feedback loop of an op amp.
- **Hybrid addition.** Two digital numbers can be applied to the inputs of two DACs, and the current produced by each number can be summed in a resistor. The measured current is proportional to the sum of inputs.
- **Visual display unit (VDU).** A DAC can be used as visual display unit (VDU) in television, in forms of scanning, and in matrix and graphic display.

Other applications of these devices are transient analysis, averaging by addition, analog delay lines, and correlation. As the technology progresses, these devices will be able to be used in new applications.

6.2 Definition of Terms

6.2.1 Resolution

Resolution is the ratio of the smallest quantity that can be accomplished with respect to the maximum obtained voltage. It is the number that the analog input voltage is divided by in every step (ADC) or the number of values for a digital input from which the analog output is constructed in DAC conversion. This number is in the order of 2^n.

Resolution can be summarized by

$$\text{Resolution} = \frac{1}{2^n}$$

where n is the number of bits. For example, the resolution of a 12 bit DAC is 1/4096 and for a 14-bit DAC it is 1/16384. Resolution may be expressed in percentage of full scale (FS). A 12-bit DAC can resolve one part in 2^{12} or one part in 4096 or 0.0245 percent of its full scale. Likewise a 10 V full-scale, 12-bit converter can resolve a 2.45 mV input change.

Resolution is one of the design parameters and does not represent the linearity or accuracy of the device.

6.2.2 Accuracy

Accuracy is usually defined as a percentage of full scale. In a DAC it is the worst-case fluctuation of the output voltage with respect to the ideal path that is drawn between zero and full-scale voltage.

The maximum conversion accuracy for a 12-bit DAC is $\pm \frac{1}{2}$ LSB (see Section 6.2.18) or ± 1 part in 2^{12+1}, that is, ± 0.0122 percent of full scale (FS), so the deviation of the accuracy of this device from 100 percent would be 99.9878 percent. In an ADC the accuracy is defined as the difference between the actual input voltage and the full-scale voltage required to produce the same binary word at the output. This difference is also called the *absolute error*, which is usually expressed as a percentage or, when the relative accuracy or error is being considered, can be expressed as a function of the quantum (details are given below). The causes of this error are nonlinearity, quantization, zero offset, gain drift, power supply sensitivity, and noise, which are fully described in the subsequent pages.

If a 12-bit ADC is said to be ± 1 LSB accurate, it means that it is equivalent to ± 0.0245 percent, which is twice as large as the minimum quantization error (0.0122 percent). Grounding is very important in the accuracy test. Figure 6.1 shows the accuracy curve for a typical converter.

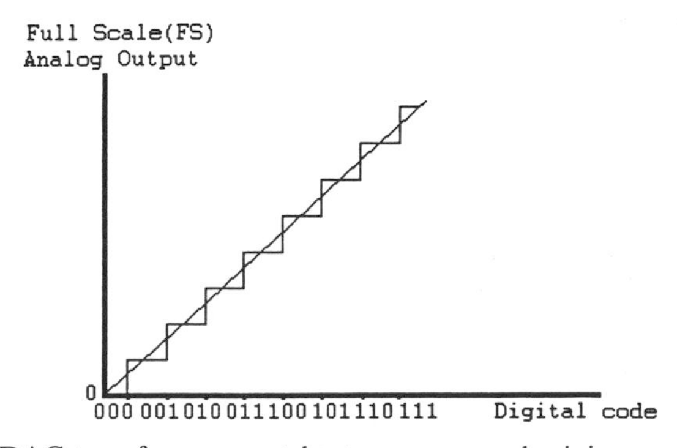

Figure 6.1 DAC transfer curves at best accuracy and minimum resolution error.

6.2.3 Quantization

Quantization is a process in which the amplitude of an analog sample is divided into a finite number of discrete levels (digits) with definite resolution by an ADC. A quantum represents one least significant bit (LSB), which is equivalent to a voltage representing the size of each cell. Quantization noise is created when the original sample values do not match the reconstructed equivalent. Also refer to Figure 6.12.

6.2.4 Quantization Error

This error is caused by converter resolution and is inherent to the principle of the ADC, an essentially nonlinear device. This error always has an absolute value of $\frac{1}{2}$ LSB regardless of the conversion method (see Figure 6.2).

Quantization error is defined as the maximum fluctuation along the straight line transfer function of a perfect A/D device. Quantization error is considered as theoretical error.

6.2.5 Full-scale Error

In D/A conversion, full-scale error is the variation of output voltage from the design specified voltage for a specific input code (usually full-scale code, see Figure 6.3). In A/D, it is the variation of the actual input voltage from the design specified input voltage for a full-scale code. It is the difference between the theoretical and actual output of an A/D when full-scale input voltage is applied. This type of error is a design fault and outside of the test engineer's control.

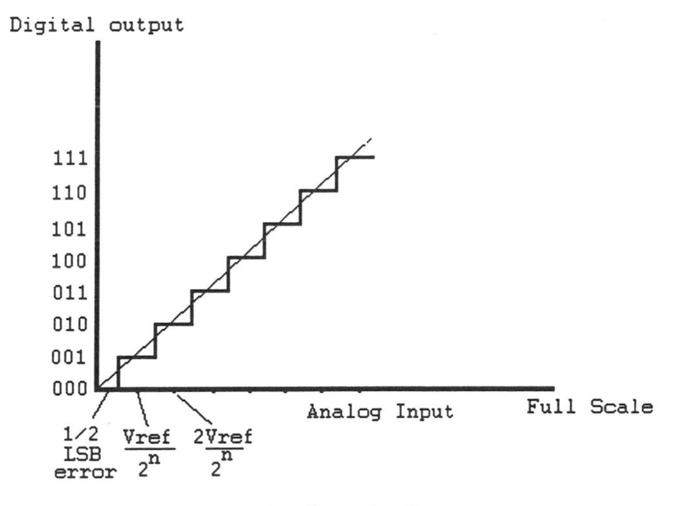

Figure 6.2 Quantization error.

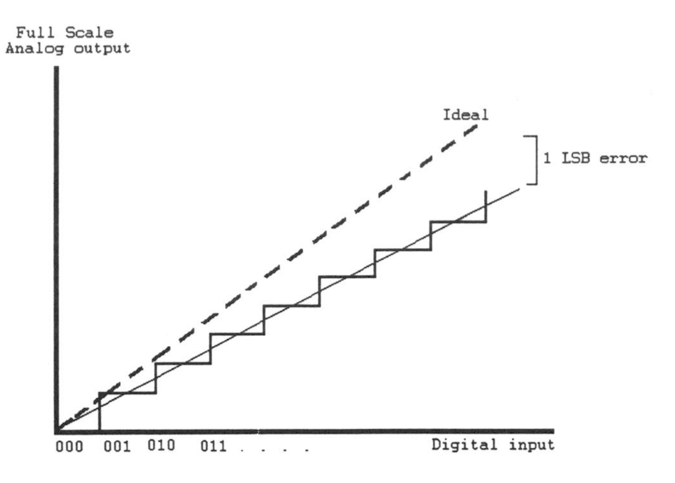

Figure 6.3 Full-scale error in D/A.

6.2.6 Gain Error

In essence, this temperature-dependent error is the difference between the theoretical and actual full scale of transition voltage. The result of this error is the rotation of the transfer function curve around its ideal location, which is expressed either in LSB terms or as a percentage of full scale (FS). The ideal location is achieved when all zero bits are used, see Figure 6.4. The greatest magnitude of this error is when the highest binary sequence is applied to the input (all ones).

6.2.7 Offset (Zero) Error

Offset error is usually expressed as a percentage of full scale or a fraction of LSB.

An ideal DAC will produce 0 V output when all inputs are zero bits. If a DAC is producing a nonzero output for this condition, it is said to have offset error.

A perfect ADC will produce zero digital output upon applying the mean analog input, which is 0 V. Any A/D device that produces this zero code upon receiving any input other than zero analog mean (which is required for an ideal ADC) is said to have offset error. Figure 6.5 shows the offset error for D/A and A/D.

6.2.8 Linearity Error

The departure of the transfer function from its ideal linearity line is a linearity error, which is very critical in converter operation. Unlike gain and offset errors, it cannot be modified. For a given input voltage the linearity error can be represented as the difference between the real curve and the theoretical one.

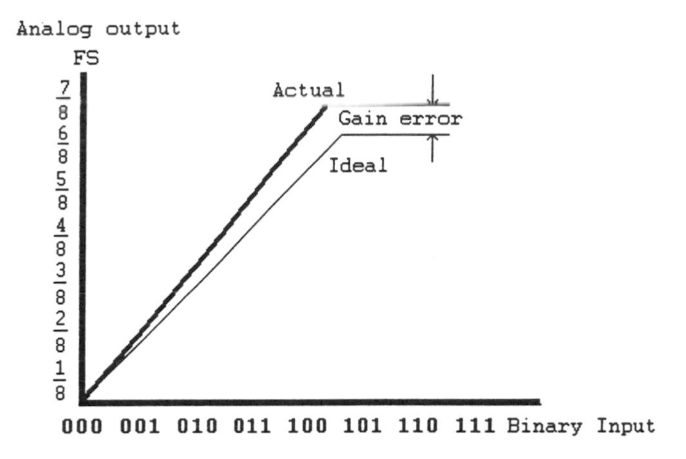

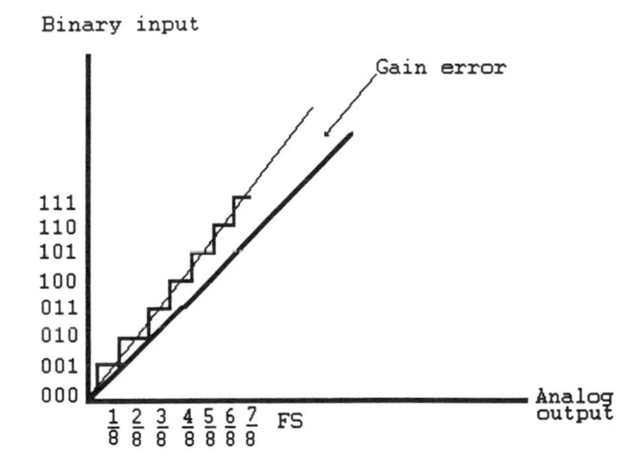

Figure 6.4 Gain error.

As shown in Figure 6.6, in an ideal DAC a straight line can be drawn to pass through all of the analog output and the same line can be drawn in an ideal ADC transfer function curve to pass through all of the analog inputs. In a nonideal converter, the transfer function departs slightly from this straight line.

6.2.9 Differential Linearity Error

The differential linearity error in a DAC is the difference between the real and ideal voltage change for any given code. For instance, assume that the quantum (Section 6.3) is 1 V. If one transition occured at 2.6 V and the next at 2.9 V, then the differential linearity error would be

$$1-(2.9-2.6)=0.7\ \text{V}$$

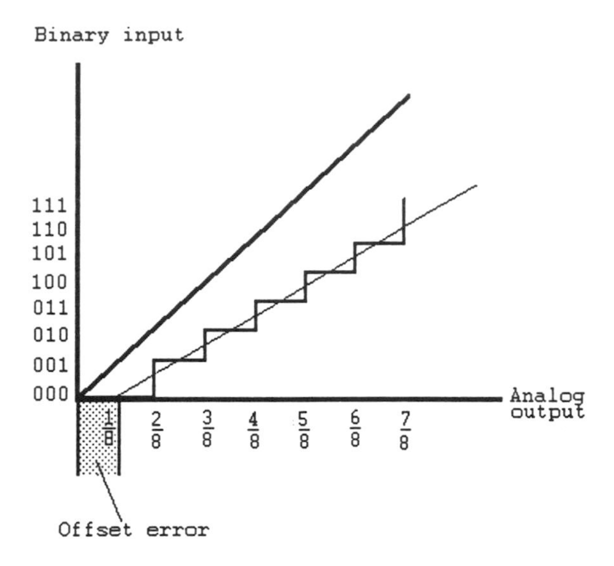

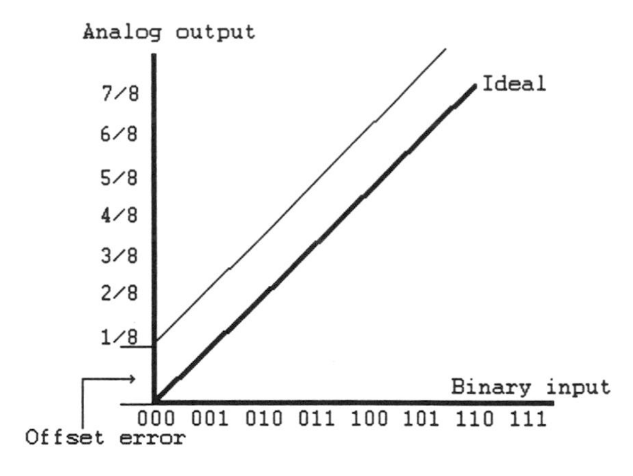

Figure 6.5 Offset error.

This difference should be less than the quantum voltage, otherwise a code will be missing, which is the worst case of departure of the actual analog output from its theoretical 1 LSB (Figure 6.7). The most effective method of testing an A/D device for the detection of the speed of the SAR or any other data conversion module, is to test the device with codes in which the largest number of bits is switched in one instant: in the case of an 8-bit converter, this would be 10000000 and 01111111. This checks the speed of the SAR against the adjacent transistors that are merging. The largest differential linearity error occurs in a D/A during the transition of major codes.

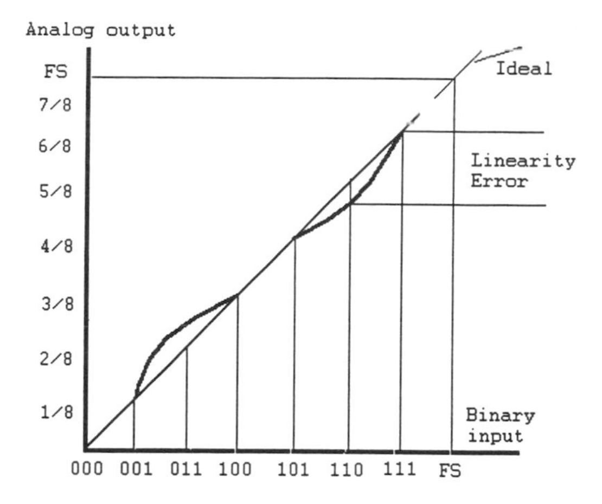

Figure 6.6 Linearity error.

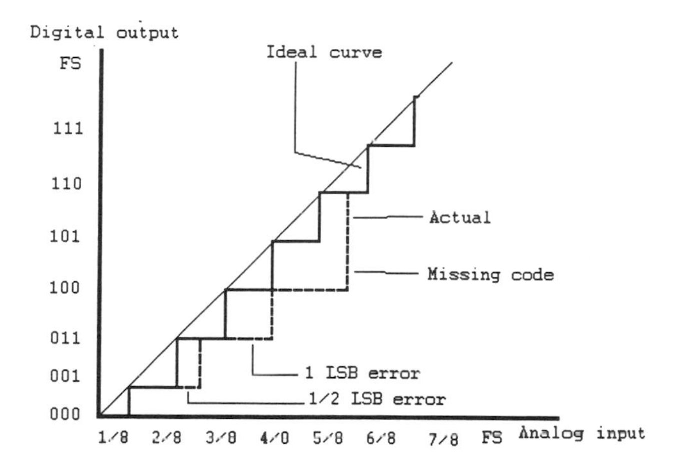

Figure 6.7 Missing codes.

6.2.10 Monotonicity

The term monotonicity is applied to D/A as well as A/D. As the binary input to the device increments, the analog output either increases or stays the same. If the differential linearity of the D/A is greater than 1 LSB, the device is said to be nonmonotonic. In other words, the output of the D/A or A/D device must be consistent with the changes (increments or decrements) at the input.

6.2.11 Settling Time

Settling time is the time the output of a D/A takes to reach and stay at $\pm\frac{1}{2}$LSB of its final value after the corresponding input digital code has been changed. The $\pm\frac{1}{2}$LSB error band is dependent on the input step change and may be defined differently for different devices. Therefore, the settling time increases as the output step change decreases.

6.2.12 Slew Rate

Slew rate is the maximum rate of change in the output voltage after a code transition. This is the characteristic limitation of the D/A output amplifier.

6.2.13 Glitches or Output Spikes

A glitch or output spike is a switching transition that appears at the output as the code transition occurs. Its value is expressed either in voltage or in current.

Sources of glitches are

- D/A input bit timing differences that result in output current spikes.
- The inability of the output amplifier to slew at high speed.

These glitches cannot be filtered due to their nonlinearity and code dependency. An effective way of suppressing them is to use a track-and-hold amplifier at the analog output of the D/A.

6.2.14 Temperature Coefficient

The performance and parameter measurement of D/A or A/D devices can be severely affected by temperature changes. The only parameters that are not affected by temperature changes are quantization and resultion. Since the operation and parameter measurement of these devices are so linearly dependent on the temperature variation, it will be necessary to add or subtract the expected changes in each parameter from the limit of that parameter for a particular temperature range.

6.2.15 Power Supply Ratio

Power supply ratio is the change in the output of a converter due to a change in the value of the power supply (see Section 5.18, "PSRR Test"). This parameter is usually expressed as a percentage of full-scale range over 1 percent change in power supply.

6.2.16 Conversion Rate

The number of measurements that are completed by a converter in a unit of time is called the conversion rate and is specified for full resolution.

6.2.17 Conversion Time

The time required to accomplish a complete conversion in the device is the conversion time. In some converters the cycle time may be longer than the conversion time due to the type of conversion module employed.

6.2.18 Least Significant Bit (LSB)

The least significant bit is the smallest analog change that can occur in a DAC output. It relates to only one bit change in the binary input. This parameter can be expressed in either voltage or current.

$$\text{Analog LSB} = \frac{\text{FSR}}{2^n}$$

where

n = number of input bits
FSR = full-scale range

6.2.19 Most Significant Bit (MSB)

The analog weight of the MSB is the largest change that can be seen at the output by switching a single input bit. Like the LSB, the value of this parameter can be expressed in voltage or current.

$$\text{Analog MSB} = \frac{\text{FSR}}{2}$$

6.2.20 Input Range (Bipolar)

The input range is centered at zero and steps in both positive or negative directions; e.g., +5 to −5 V. In bipolar mode, the converter is capable of operating with either a positive or a negative input voltage.

6.2.21 Input Range (Unipolar)

Unlike the bipolar input range, this input starts from zero and steps in either a positive or a negative direction, producing single-polarity input voltage.

In the unipolar mode, the converter is only capable of converting one type of input voltage, positive or negative. Converter devices have a variety of inputs, which range over unipolar, bipolar, or the combination of both. For each type of input the expected outputs are predefined.

The input range details for any device family are outlined in the manufacturer's data book. A test program for this family of devices must include all the ranges.

6.3 Operation of a 4-Bit Converter

Since understanding the operation of an ADC is most critical for the test engineer, the operation of a 4-bit ADC is described here in detail, including some of the terms that have been mentioned previously.

Four-bit binary digits can have only 16 different combinations beginning at $(0000)_2$ and ending at $(1111)_2$ or $(2^4=16)_{10}$. If the input range of this device is from 0 V to +10 V (unipolar), then the graphic representation of the device function is as shown in Figure 6.8.

An analog signal can be partitioned into an infinite number of segments. From the above figure we can conclude the following.

1. Each of the above cells is one least significant bit (LSB) wide. Since any analog input voltage can have infinite resolution, the LSB can be large or small depending on the number of converter bits and the operating voltage range.
2. If a voltage is applied to the input, we should expect a sequence of digital output. For, say, 4 V applied to the input, the corresponding digital output (predefined) should be the sequence $(0110)_2$, which is the content of cell #7.
3. Conversely, we can find the value of the input by looking at the digital output. For instance, the sequence $(0011)_2$ at the output represents a voltage between 1.875 V and 2.5 V applied to the input. In our case, 1 LSB = 0.625 V, which is also called the *quantum* (10 V/16 bits = 0.625 V per bit). As mentioned before, quantization is the process (in ADC) by which a fraction of a volt of analog input of essentially infinite resolution can be converted to a digital code with definite resolution.
4. If the voltage of the center of each of the above cells is assigned definite output digits, then the ADC can function as a voltmeter with an accuracy of

$$\pm\tfrac{1}{2}\,\text{LSB} = \frac{0.625}{2} = \pm 0.3125\ \text{V}$$

This is called *quantization error*. Obviously a higher number of ADC bits and a lower input voltage range produce a more accurate voltmeter.

Figure 6.9 shows a 4-bit bipolar DAC. Relays K1 through K4 represent binary bits and K5 selects the polarity of the output voltage. The output ranges

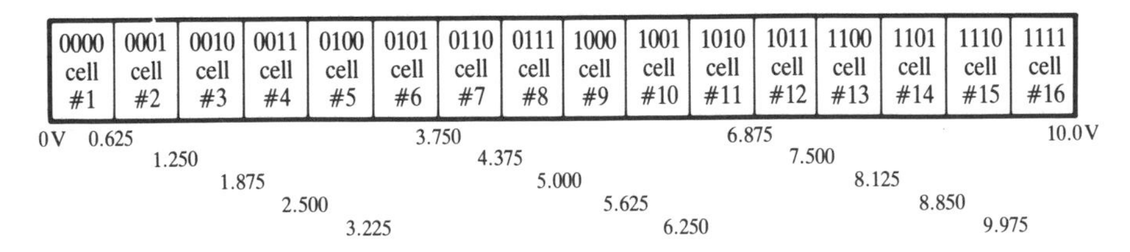

Figure 6.8 A graphic representation of a 4-bit A/D converter.

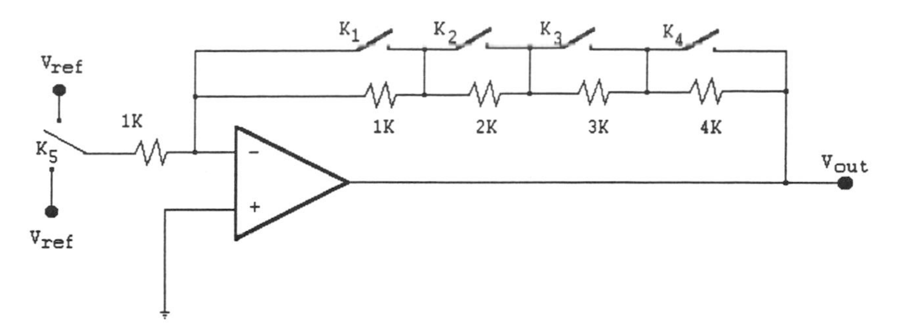

Figure 6.9 Representation of a 4-bit DAC.

from 0 V to ±10 V depending on the K5 relay status. Each step increments or decrements by ±0.625 V. This DAC uses the 2's complement to generate the negative output. The 2's complement of a binary number is produced by converting every bit of the binary train and then adding binary 1 to it. Depending on the accuracy of the input voltage and resistors used, the output can be fairly accurate. The accuracy of the op amp output does not indicate that the generated output stays the same until it is used in another stage of the system, because there is a constant voltage drop along any conductor.

The 2's complementation is performed by an SAR internal to the device. There is a pin on every DAC that is designated for this operation. The time required for the SAR to complement one binary bit is dependent on the total number of bits in the input train. The process is controlled by the internal clock and it usually takes one clock pulse to find each bit in a train of binary bits. The system requires $n+1$ clock cycles to convert n binary digits. (Refer to Figure 6.10.)

6.4 Conversion Operation

The following operations are carried out during conversion:

- Sampling (by the sample and hold circuit)
- Quantization
- A/D or D/A conversion
- Recovery

6.4.1 Sampling

In the process of *sampling*, a continuous function will be converted into a function with a number of cut-out segments. The result is a discontinuous function (Figure 6.11). This operation is carried out at a rate that is solely dependent on the shape of the input signal. (See Chapter 7 for in-depth detail.)

−15	0 0 0 1	−10.00	v
−14	0 0 1 0	−8.85	v
−13	0 0 1 1	−8.125	v
−12	0 1 0 0	−7.5	v
−11	0 1 0 1	−6.875	v
−10	0 1 1 0	−6.25	v
−9	0 1 1 1	−5.625	v
−8	1 0 0 0	−5.00	v
−7	1 0 0 1	−4.375	v
−6	1 0 1 0	−3.75	v
−5	1 0 1 1	−3.225	v
−4	1 1 0 0	−2.5	v
−3	1 1 0 1	−1.875	v
−2	1 1 1 0	−1.25	v
−1	1 1 1 1	−0.625	v
	0 0 0 0	−0.00	v
			v
	0 0 0 0	+0.00	v
	0 0 0 1	+0.625	v
	0 0 1 0	+1.25	v
	0 0 1 1	+1.875	v
	0 1 0 0	+2.5	v
	0 1 0 1	+3.225	v
	0 1 1 0	+3.75	v
	0 1 1 1	+4.375	v
	1 0 0 0	+5.00	v
	1 0 0 1	+5.625	v
	1 0 1 0	+6.25	v
	1 0 1 1	+6.875	v
	1 1 0 0	+7.5	v
	1 1 0 1	+8.125	v
	1 1 1 0	+8.85	v
	1 1 1 1	+10.00	v

Figure 6.10 Bipolar input and complementation.

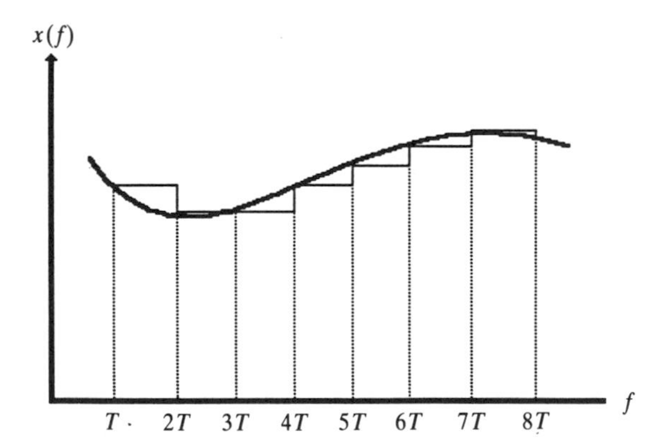

Figure 6.11 A sampled analog signal.

6.4.2 Recovery

The sample that has been converted either from digital to analog or vice versa should be recovered. The maximum frequency of the sample and the sampling rate are two essential parameters that are required for this purpose. The information

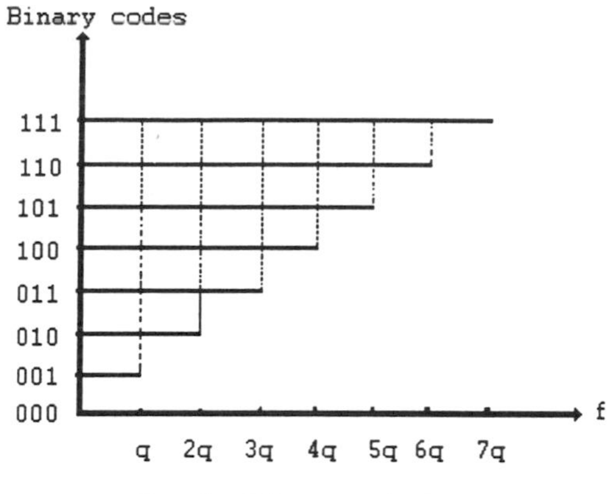

Encoding follows quantization

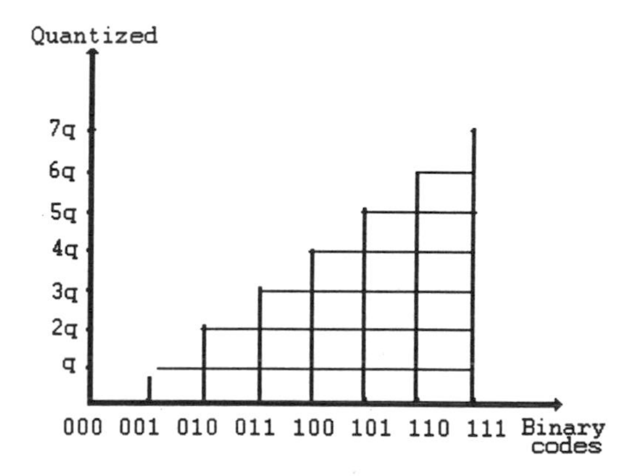

Figure 6.12 D/A conversion.

is passed through a low-pass filter with the cutoff frequency and the output of the filter will be the original signal.

6.5 A/D and D/A Test Description

Most parametric tests that are done on these devices are the same as those outlined in Chapter 3. The only differences are in those that make the operation of this group of devices unique.

The biggest problem in A/D testing is the many variations to be tested. The output of the A/D device is in the form of predefined binary codes, each of which represents and defines the analog input voltage. The outputs can be read in a variety of modes. These output binary codes can be read in 1's or 2's complement, binary coded decimal, sign and magnitude, or a combination of any of these. Obviously a 12-bit A/D converter has 4095 codes to test.

A data acquisition system including an 8-channel multiplexer (MUX) and a microprocessor-compatible control logic device is chosen as an example since it will provide a rather detailed insight into how an ADC may be tested and the reasoning and intention of each test. The layout of this converter is given in Figure 6.13. The following semiconductor manufacturers fabricate such devices with almost the same function.

Manufacturer	*Device ID*	*Specification*	
Samsung	KAD809	CMOS	8-ch. MUX, SAR
Micro Linear	ML2258	CMOS	8-ch. MUX, SAR
Texas Instrument	ADC0809	CMOS	8-ch. MUX, SAR
National Semiconductor Corp.	ADC809	CMOS	8-ch. MUX, SAR
Motorola	MC14442	CMOS	8-ch. MUX, SAR

This typical ADC has an internal 8-bit MUX. The conversion is performed by SAR. The 8-bit MUX can access any of the eight input signals. Figure 6.13 gives the pin definition.

This chapter outlines testing procedures for all the parameters that adequately guarantee the functionality of the device as suggested by most manufacturers. The method that is outlined here is only one way of testing each parameter. Better, quicker, or more efficient methods depend solely on the creativity of the test engineer and, of course, the capability of the test system.

Figure 6.14 represents the data sheets from National Semiconductor Corporation for the ADC809 8-bit microprocessor-compatible converters with

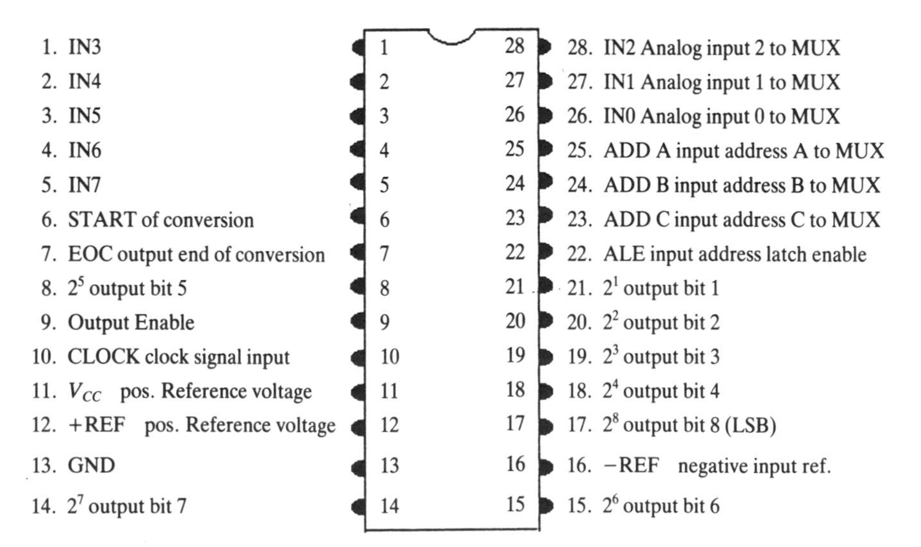

Figure 6.13 Pin description of ADC809.

ADC0808/ADC0809 8-Bit μP Compatible A/D Converters with 8-Channel Multiplexer

General Description

The ADC0808, ADC0809 data acquisition component is a monolithic CMOS device with an 8-bit analog-to-digital converter, 8-channel multiplexer and microprocessor compatible control logic. The 8-bit A/D converter uses successive approximation as the conversion technique. The converter features a high impedance chopper stabilized comparator, a 256R voltage divider with analog switch tree and a successive approximation register. The 8-channel multiplexer can directly access any of 8-single-ended analog signals.

The device eliminates the need for external zero and full-scale adjustments. Easy interfacing to microprocessors is provided by the latched and decoded multiplexer address inputs and latched TTL TRI-STATE® outputs.

The design of the ADC0808, ADC0809 has been optimized by incorporating the most desirable aspects of several A/D conversion techniques. The ADC0808, ADC0809 offers high speed, high accuracy, minimal temperature dependence, excellent long-term accuracy and repeatability, and consumes minimal power. These features make this device ideally suited to applications from process and machine control to consumer and automotive applications. For 16-channel multiplexer with common output (sample/hold port) see ADC0816 data sheet. (See AN-247 for more information.)

Features

- Easy interface to all microprocessors
- Operates ratiometrically or with 5 V_{DC} or analog span adjusted voltage reference
- No zero or full-scale adjust required
- 8-channel multiplexer with address logic
- 0V to 5V input range with single 5V power supply
- Outputs meet TTL voltage level specifications
- Standard hermetic or molded 28-pin DIP package
- 28-pin molded chip carrier package
- ADC0808 equivalent to MM74C949
- ADC0809 equivalent to MM74C949-1

Key Specifications

■ Resolution	8 Bits
■ Total Unadjusted Error	±½ LSB and ±1 LSB
■ Single Supply	5 V_{DC}
■ Low Power	15 mW
■ Conversion Time	100 μs

Block Diagram

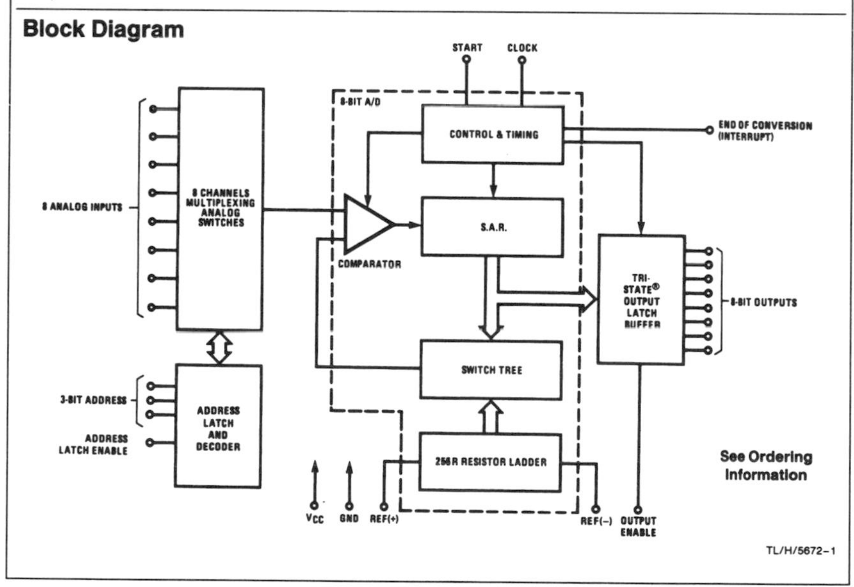

See Ordering Information

TL/H/5672-1

Figure 6.14 Complete data sheets for an analog-to-digital converter (ADC0809) chip manufactured by National Semiconductor Corporation. Information contained in data sheets is usually adequate for test, product, and application engineers as well as the for the user to become familiar with the operation and physical characteristics of the device. (Data sheets are generated by manufacturers.) (Reprinted with permission of National Semiconducter Corp.)

Absolute Maximum Ratings (Notes 1 & 2)

If Military/Aerospace specified devices are required, please contact the National Semiconductor Sales Office/Distributors for availability and specifications.

Supply Voltage (V_{CC}) (Note 3)	6.5V
Voltage at Any Pin Except Control Inputs	−0.3V to (V_{CC}+0.3V)
Voltage at Control Inputs (START, OE, CLOCK, ALE, ADD A, ADD B, ADD C)	−0.3V to +15V
Storage Temperature Range	−65°C to +150°C
Package Dissipation at T_A=25°C	875 mW
Lead Temp. (Soldering, 10 seconds)	
Dual-In-Line Package (plastic)	260°C
Dual-In-Line Package (ceramic)	300°C
Molded Chip Carrier Package	
Vapor Phase (60 seconds)	215°C
Infrared (15 seconds)	220°C
ESD Susceptibility (Note 11)	400V

Operating Conditions (Notes 1 & 2)

Temperature Range (Note 1)	$T_{MIN} \le T_A \le T_{MAX}$
ADC0808CJ	$-55°C \le T_A \le +125°C$
ADC0808CCJ, ADC0808CCN, ADC0809CCN	$-40°C \le T_A \le +85°C$
ADC0808CCV, ADC0809CCV	$-40°C \le T_A \le +85°C$
Range of V_{CC} (Note 1)	4.5 V_{DC} to 6.0 V_{DC}

Electrical Characteristics

Converter Specifications: V_{CC}=5 V_{DC}=V_{REF+}, $V_{REF(-)}$=GND, $T_{MIN} \le T_A \le T_{MAX}$ and f_{CLK}=640 kHz unless otherwise stated.

Symbol	Parameter	Conditions	Min	Typ	Max	Units
	ADC0808 Total Unadjusted Error (Note 5)	25°C T_{MIN} to T_{MAX}			±½ ±¾	LSB LSB
	ADC0809 Total Unadjusted Error (Note 5)	0°C to 70°C T_{MIN} to T_{MAX}			±1 ±1¼	LSB LSB
	Input Resistance	From Ref(+) to Ref(−)	1.0	2.5		kΩ
	Analog Input Voltage Range	(Note 4) V(+) or V(−)	GND−0.10		V_{CC}+0.10	V_{DC}
$V_{REF(+)}$	Voltage, Top of Ladder	Measured at Ref(+)		V_{CC}	V_{CC}+0.1	V
$\frac{V_{REF(+)}+V_{REF(-)}}{2}$	Voltage, Center of Ladder		V_{CC}/2-0.1	V_{CC}/2	V_{CC}/2+0.1	V
$V_{REF(-)}$	Voltage, Bottom of Ladder	Measured at Ref(−)	−0.1	0		V
I_{IN}	Comparator Input Current	f_c=640 kHz, (Note 6)	−2	±0.5	2	μA

Electrical Characteristics

Digital Levels and DC Specifications: ADC0808CJ 4.5V≤V_{CC}≤5.5V, −55°C≤T_A≤+125°C unless otherwise noted ADC0808CCJ, ADC0808CCN, ADC0808CCV, ADC0809CCN and ADC0809CCV, 4.75≤V_{CC}≤5.25V, −40°C≤T_A≤+85°C unless otherwise noted

Symbol	Parameter	Conditions	Min	Typ	Max	Units
ANALOG MULTIPLEXER						
$I_{OFF(+)}$	OFF Channel Leakage Current	V_{CC}=5V, V_{IN}=5V, T_A=25°C T_{MIN} to T_{MAX}		10	200 1.0	nA μA
$I_{OFF(-)}$	OFF Channel Leakage Current	V_{CC}=5V, V_{IN}=0, T_A=25°C T_{MIN} to T_{MAX}	−200 −1.0	−10		nA μA

Figure 6.14—*continued*

Electrical Characteristics (Continued)

Digital Levels and DC Specifications: ADC0808CJ $4.5V \leq V_{CC} \leq 5.5V$, $-55°C \leq T_A \leq +125°C$ unless otherwise noted ADC0808CCJ, ADC0808CCN, ADC0808CCV, ADC0809CCN and ADC0809CCV, $4.75 \leq V_{CC} \leq 5.25V$, $-40°C \leq T_A \leq +85°C$ unless otherwise noted

Symbol	Parameter	Conditions	Min	Typ	Max	Units
CONTROL INPUTS						
$V_{IN(1)}$	Logical "1" Input Voltage		$V_{CC}-1.5$			V
$V_{IN(0)}$	Logical "0" Input Voltage				1.5	V
$I_{IN(1)}$	Logical "1" Input Current (The Control Inputs)	$V_{IN}=15V$			1.0	μA
$I_{IN(0)}$	Logical "0" Input Current (The Control Inputs)	$V_{IN}=0$	−1.0			μA
I_{CC}	Supply Current	$f_{CLK}=640$ kHz		0.3	3.0	mA
DATA OUTPUTS AND EOC (INTERRUPT)						
$V_{OUT(1)}$	Logical "1" Output Voltage	$I_O=-360$ μA	$V_{CC}-0.4$			V
$V_{OUT(0)}$	Logical "0" Output Voltage	$I_O=1.6$ mA			0.45	V
$V_{OUT(0)}$	Logical "0" Output Voltage EOC	$I_O=1.2$ mA			0.45	V
I_{OUT}	TRI-STATE Output Current	$V_O=5V$ $V_O=0$	 −3		3	μA μA

Electrical Characteristics

Timing Specifications $V_{CC}=V_{REF(+)}=5V$, $V_{REF(-)}=GND$, $t_r=t_f=20$ ns and $T_A=25°C$ unless otherwise noted.

Symbol	Parameter	Conditions	Min	Typ	Max	Units
t_{WS}	Minimum Start Pulse Width	*(Figure 5)*		100	200	ns
t_{WALE}	Minimum ALE Pulse Width	*(Figure 5)*		100	200	ns
t_s	Minimum Address Set-Up Time	*(Figure 5)*		25	50	ns
t_H	Minimum Address Hold Time	*(Figure 5)*		25	50	ns
t_D	Analog MUX Delay Time From ALE	$R_S=0Ω$ *(Figure 5)*		1	2.5	μS
t_{H1}, t_{H0}	OE Control to Q Logic State	$C_L=50$ pF, $R_L=10k$ *(Figure 8)*		125	250	ns
t_{1H}, t_{0H}	OE Control to Hi-Z	$C_L=10$ pF, $R_L=10k$ *(Figure 8)*		125	250	ns
t_c	Conversion Time	$f_c=640$ kHz, *(Figure 5)* (Note 7)	90	100	116	μS
f_c	Clock Frequency		10	640	1280	kHz
t_{EOC}	EOC Delay Time	*(Figure 5)*	0		8+2 μS	Clock Periods
C_{IN}	Input Capacitance	At Control Inputs		10	15	pF
C_{OUT}	TRI-STATE Output Capacitance	At TRI-STATE Outputs, (Note 12)		10	15	pF

Note 1: Absolute Maximum Ratings indicate limits beyond which damage to the device may occur. DC and AC electrical specifications do not apply when operating the device beyond its specified operating conditions.

Note 2: All voltages are measured with respect to GND, unless othewise specified.

Note 3: A zener diode exists, internally, from V_{CC} to GND and has a typical breakdown voltage of 7 V_{DC}.

Note 4: Two on-chip diodes are tied to each analog input which will forward conduct for analog input voltages one diode drop below ground or one diode drop greater than the V_{CC}n supply. The spec allows 100 mV forward bias of either diode. This means that as long as the analog V_{IN} does not exceed the supply voltage by more than 100 mV, the output code will be correct. To achieve an absolute $0V_{DC}$ to $5V_{DC}$ input voltage range will therefore require a minimum supply voltage of 4.900 V_{DC} over temperature variations, initial tolerance and loading.

Note 5: Total unadjusted error includes offset, full-scale, linearity, and multiplexer errors. See *Figure 3*. None of these A/Ds requires a zero or full-scale adjust. However, if an all zero code is desired for an analog input other than 0.0V, or if a narrow full-scale span exists (for example: 0.5V to 4.5V full-scale) the reference voltages can be adjusted to achieve this. See *Figure 13*.

Note 6: Comparator input current is a bias current into or out of the chopper stabilized comparator. The bias current varies directly with clock frequency and has little temperature dependence *(Figure 6)*. See paragraph 4.0.

Note 7: The outputs of the data register are updated one clock cycle before the rising edge of EOC.

Note 8: Human body model, 100 pF discharged through a 1.5 kΩ resistor.

Figure 6.14—*continued*

Functional Description

Multiplexer. The device contains an 8-channel single-ended analog signal multiplexer. A particular input channel is selected by using the address decoder. Table I shows the input states for the address lines to select any channel. The address is latched into the decoder on the low-to-high transition of the address latch enable signal.

TABLE I

SELECTED ANALOG CHANNEL	ADDRESS LINE		
	C	B	A
IN0	L	L	L
IN1	L	L	H
IN2	L	H	L
IN3	L	H	H
IN4	H	L	L
IN5	H	L	H
IN6	H	H	L
IN7	H	H	H

CONVERTER CHARACTERISTICS

The Converter

The heart of this single chip data acquisition system is its 8-bit analog-to-digital converter. The converter is designed to give fast, accurate, and repeatable conversions over a wide range of temperatures. The converter is partitioned into 3 major sections: the 256R ladder network, the successive approximation register, and the comparator. The converter's digital outputs are positive true.

The 256R ladder network approach *(Figure 1)* was chosen over the conventional R/2R ladder because of its inherent monotonicity, which guarantees no missing digital codes. Monotonicity is particularly important in closed loop feedback control systems. A non-monotonic relationship can cause oscillations that will be catastrophic for the system. Additionally, the 256R network does not cause load variations on the reference voltage.

The bottom resistor and the top resistor of the ladder network in *Figure 1* are not the same value as the remainder of the network. The difference in these resistors causes the output characteristic to be symmetrical with the zero and full-scale points of the transfer curve. The first output transition occurs when the analog signal has reached $+\frac{1}{2}$ LSB and succeeding output transitions occur every 1 LSB later up to full-scale.

The successive approximation register (SAR) performs 8 iterations to approximate the input voltage. For any SAR type converter, n-iterations are required for an n-bit converter. *Figure 2* shows a typical example of a 3-bit converter. In the ADC0808, ADC0809, the approximation technique is extended to 8 bits using the 256R network.

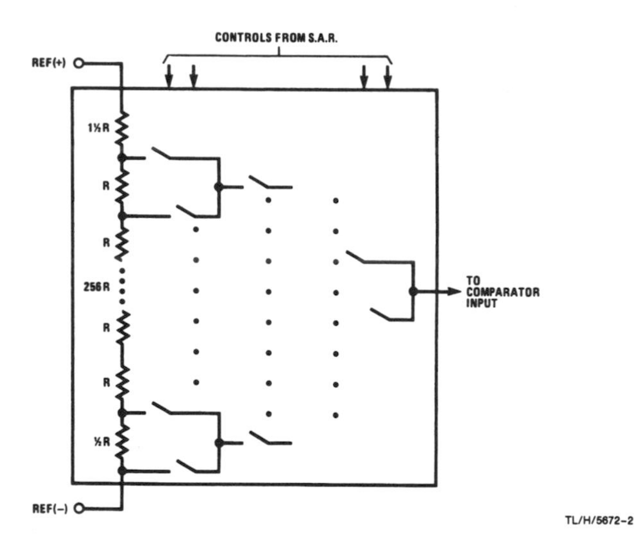

FIGURE 1. Resistor Ladder and Switch Tree

Figure 6.14—*continued*

Functional Description (Continued)

The A/D converter's successive approximation register (SAR) is reset on the positive edge of the start conversion (SC) pulse. The conversion is begun on the falling edge of the start conversion pulse. A conversion in process will be interrupted by receipt of a new start conversion pulse. Continuous conversion may be accomplished by tying the end-of-conversion (EOC) output to the SC input. If used in this mode, an external start conversion pulse should be applied after power up. End-of-conversion will go low between 0 and 8 clock pulses after the rising edge of start conversion.

The most important section of the A/D converter is the comparator. It is this section which is responsible for the ultimate accuracy of the entire converter. It is also the comparator drift which has the greatest influence on the repeatability of the device. A chopper-stabilized comparator provides the most effective method of satisfying all the converter requirements.

The chopper-stabilized comparator converts the DC input signal into an AC signal. This signal is then fed throught a high gain AC amplifier and has the DC level restored. This technique limits the drift component of the amplifier since the drift is a DC component which is not passed by the AC amplifier. This makes the entire A/D converter extremely insensitive to temperature, long term drift and input offset errors.

Figure 4 shows a typical error curve for the ADC0808 as measured using the procedures outlined in AN-179.

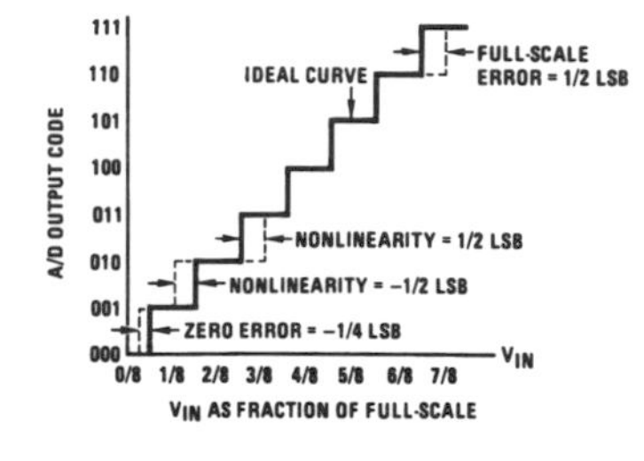

FIGURE 2. 3-Bit A/D Transfer Curve

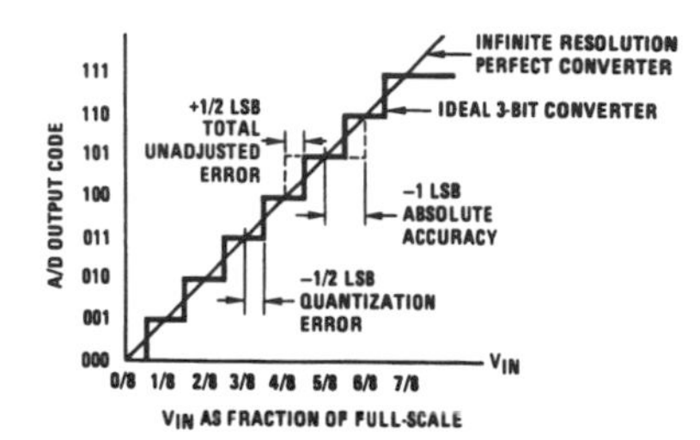

FIGURE 3. 3-Bit A/D Absolute Accuracy Curve

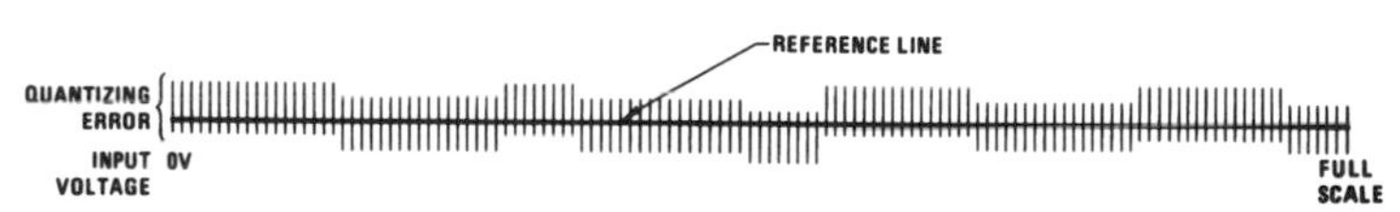

TL/H/5672–3

FIGURE 4. Typical Error Curve

Figure 6.14—*continued*

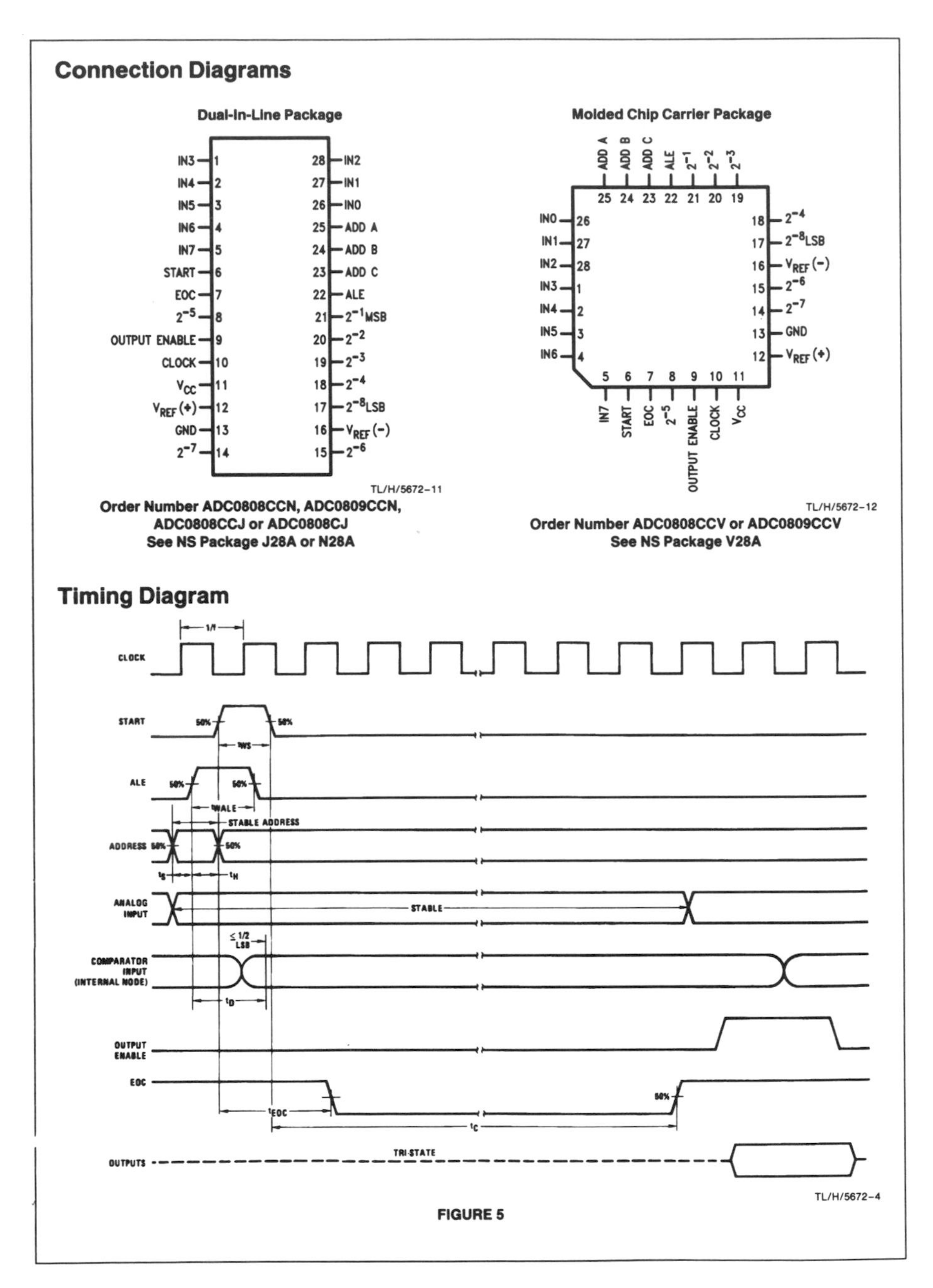

Figure 6.14—*continued*

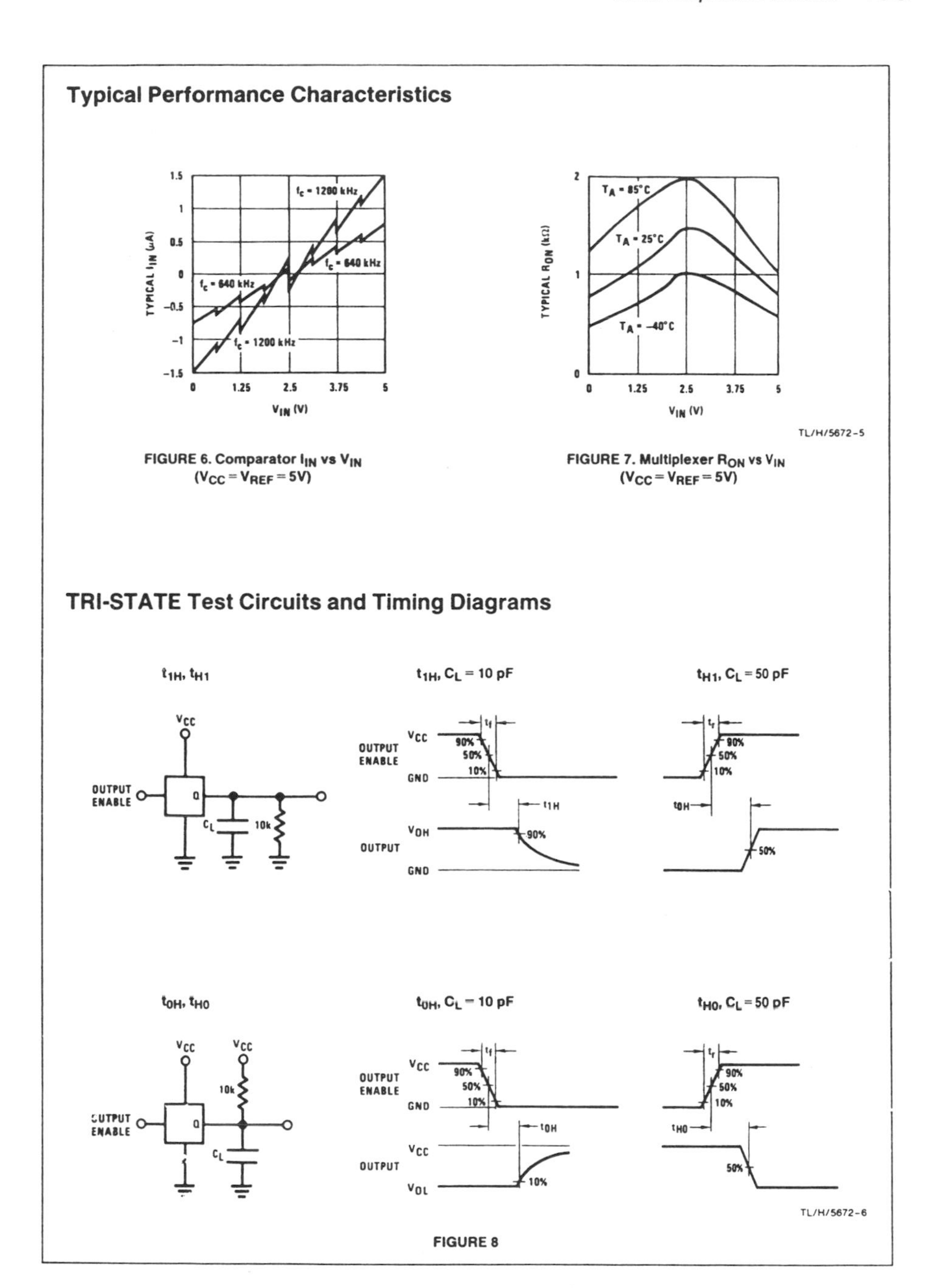

Figure 6.14—*continued*

Applications Information

OPERATION

1.0 RATIOMETRIC CONVERSION

The ADC0808, ADC0809 is designed as a complete Data Acquisition System (DAS) for ratiometric conversion systems. In ratiometric systems, the physical variable being measured is expressed as a percentage of full-scale which is not necessarily related to an absolute standard. The voltage input to the ADC0808 is expressed by the equation

$$\frac{V_{IN}}{V_{fs}-V_{Z}}=\frac{D_X}{D_{MAX}-D_{MIN}} \quad (1)$$

V_{IN} = Input voltage into the ADC0808

V_{fs} = Full-scale voltage

V_Z = Zero voltage

D_X = Data point being measured

D_{MAX} = Maximum data limit

D_{MIN} = Minimum data limit

A good example of a ratiometric transducer is a potentiometer used as a position sensor. The position of the wiper is directly proportional to the output voltage which is a ratio of the full-scale voltage across it. Since the data is represented as a proportion of full-scale, reference requirements are greatly reduced, eliminating a large source of error and cost for many applications. A major advantage of the ADC0808, ADC0809 is that the input voltage range is equal to the supply range so the transducers can be connected directly across the supply and their outputs connected directly into the multiplexer inputs, *(Figure 9)*.

Ratiometric transducers such as potentiometers, strain gauges, thermistor bridges, pressure transducers, etc., are suitable for measuring proportional relationships; however, many types of measurements must be referred to an absolute standard such as voltage or current. This means a system reference must be used which relates the full-scale voltage to the standard volt. For example, if $V_{CC}=V_{REF}=5.12V$, then the full-scale range is divided into 256 standard steps. The smallest standard step is 1 LSB which is then 20 mV.

2.0 RESISTOR LADDER LIMITATIONS

The voltages from the resistor ladder are compared to the selected into 8 times in a conversion. These voltages are coupled to the comparator via an analog switch tree which is referenced to the supply. The voltages at the top, center and bottom of the ladder must be controlled to maintain proper operation.

The top of the ladder, Ref(+), should not be more positive than the supply, and the bottom of the ladder, Ref(−), should not be more negative than ground. The center of the ladder voltage must also be near the center of the supply because the analog switch tree changes from N-channel switches to P-channel switches. These limitations are automatically satisfied in ratiometric systems and can be easily met in ground referenced systems.

Figure 10 shows a ground referenced system with a separate supply and reference. In this system, the supply must be trimmed to match the reference voltage. For instance, if a 5.12V is used, the supply should be adjusted to the same voltage within 0.1V.

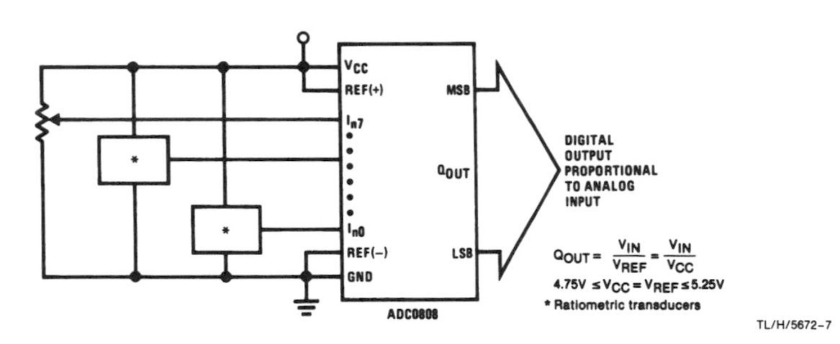

FIGURE 9. Ratiometric Conversion System

Figure 6.14—*continued*

Applications Information (Continued)

The ADC0808 needs less than a milliamp of supply current so developing the supply from the reference is readily accomplished. In *Figure 11* a ground referenced system is shown which generates the supply from the reference. The buffer shown can be an op amp of sufficient drive to supply the milliamp of supply current and the desired bus drive, or if a capacitive bus is driven by the outputs a large capacitor will supply the transient supply current as seen in *Figure 12.* The LM301 is overcompensated to insure stability when loaded by the 10 μF output capacitor.

The top and bottom ladder voltages cannot exceed V_{CC} and ground, respectively, but they can be symmetrically less than V_{CC} and greater than ground. The center of the ladder voltage should always be near the center of the supply. The sensitivity of the converter can be increased, (i.e., size of the LSB steps decreased) by using a symmetrical reference system. In *Figure 13*, a 2.5V reference is symmetrically centered about $V_{CC}/2$ since the same current flows in identical resistors. This system with a 2.5V reference allows the LSB bit to be half the size of a 5V reference system.

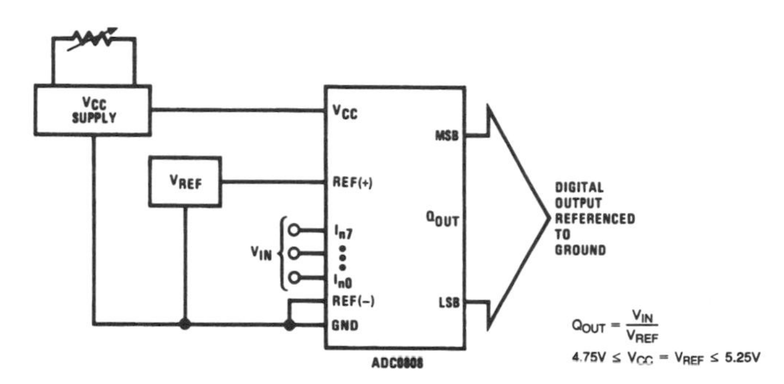

$$Q_{OUT} = \frac{V_{IN}}{V_{REF}}$$

$4.75V \le V_{CC} = V_{REF} \le 5.25V$

FIGURE 10. Ground Referenced Conversion System Using Trimmed Supply

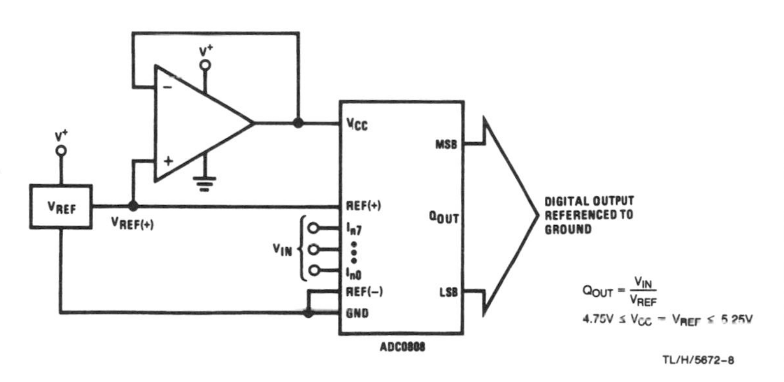

$$Q_{OUT} = \frac{V_{IN}}{V_{REF}}$$

$4.75V \le V_{CC} = V_{REF} \le 5.25V$

TL/H/5672–8

FIGURE 11: Ground Referenced Conversion System with Reference Generating V_{CC} Supply

Figure 6.14—*continued*

Typical Application

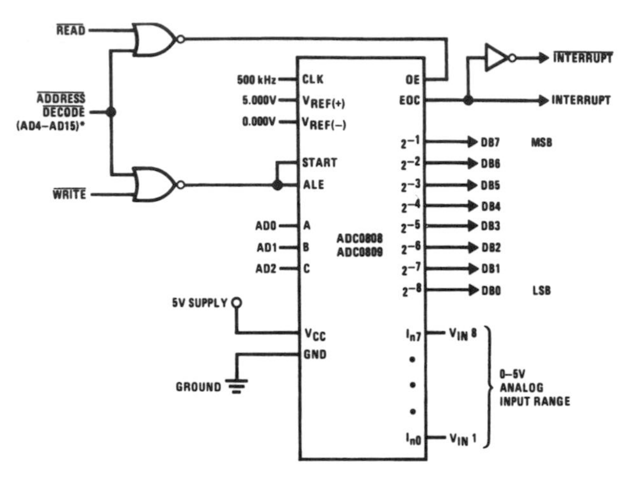

*Address latches needed for 8085 and SC/MP interfacing the ADC0808 to a microprocessor

MICROPROCESSOR INTERFACE TABLE

PROCESSOR	$\overline{\text{READ}}$	$\overline{\text{WRITE}}$	INTERRUPT (COMMENT)
8080	$\overline{\text{MEMR}}$	$\overline{\text{MEMW}}$	INTR (Thru RST Circuit)
8085	$\overline{\text{RD}}$	$\overline{\text{WR}}$	INTR (Thru RST Circuit)
Z-80	$\overline{\text{RD}}$	$\overline{\text{WR}}$	$\overline{\text{INT}}$ (Thru RST Circuit, Mode 0)
SC/MP	NRDS	NWDS	SA (Thru Sense A)
6800	VMA•ϕ2•R/W	VMA•ϕ•$\overline{\text{R/W}}$	$\overline{\text{IRQA}}$ or $\overline{\text{IRQB}}$ (Thru PIA)

Ordering Information

TEMPERATURE RANGE		−40°C to +85°C			−55°C to +125°C
Error	±½ LSB Unadjusted	ADC0808CCN	ADC0808CCV	ADC0808CCJ	ADC0808CJ
	±1 LSB Unadjusted	ADC0809CCN	ADC0809CCV		
Package Outline		N28A Molded DIP	V28A Molded Chip Carrier	J28A Ceramic DIP	J28A Ceramic DIP

Figure 6.14—*continued*

Applications Information (Continued)

FIGURE 12. Typical Reference and Supply Circuit

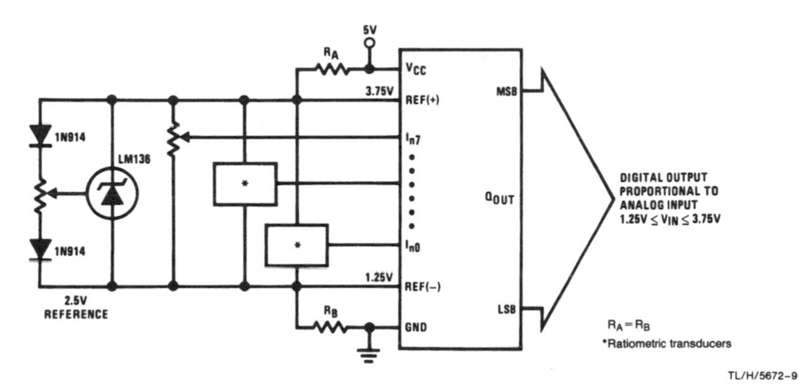

FIGURE 13. Symmetrically Centered Reference

3.0 CONVERTER EQUATIONS

The transition between adjacent codes N and N+1 is given by:

$$V_{IN} = \left\{ (V_{REF(+)} - V_{REF(-)}) \left[\frac{N}{256} + \frac{1}{512} \right] \pm V_{TUE} \right\} + V_{REF(-)} \quad (2)$$

The center of an output code N is given by:

$$V_{IN} \left\{ (V_{REF(+)} - V_{REF(-)}) \left[\frac{N}{256} \right] \pm V_{TUE} \right\} + V_{REF(-)} \quad (3)$$

The output code N for an arbitrary input are the integers within the range:

$$N = \frac{V_{IN} - V_{REF(-)}}{V_{REF(+)} - V_{REF(-)}} \times 256 \pm \text{Absolute Accuracy} \quad (4)$$

where: V_{IN} = Voltage at comparator input

$V_{REF(+)}$ = Voltage at Ref(+)

$V_{REF(-)}$ = Voltage at Ref(−)

V_{TUE} = Total unadjusted error voltage (typically $V_{REF(+)} \div 512$)

4.0 ANALOG COMPARATOR INPUTS

The dynamic comparator input current is caused by the periodic switching of on-chip stray capacitances. These are connected alternately to the output of the resistor ladder/switch tree network and to the comparator input as part of the operation of the chopper stabilized comparator.

The average value of the comparator input current varies directly with clock frequency and with V_{IN} as shown in *Figure 6*.

If no filter capacitors are used at the analog inputs and the signal source impedances are low, the comparator input current should not introduce converter errors, as the transient created by the capacitance discharge will die out before the comparator output is strobed.

If input filter capacitors are desired for noise reduction and signal conditioning they will tend to average out the dynamic comparator input current. It will then take on the characteristics of a DC bias current whose effect can be predicted conventionally.

Figure 6.14—*continued*

8-channel multiplexer. The eight channel MUX in this chip can have direct access to any of the eight analog input signals. This device can easily be interfaced to a microprocessor by the latched and decoded address inputs and latched three-state (high-impedance) output.

From a design point of view, this converter consists of three main segments:

- **256R ladder network.** The top and bottom resistors of the ladder have different values from the ones in between. This results in symmetrical output from zero to full-scale points on the transfer curve. The first output transition takes place when the analog signal reaches $\frac{1}{2}$ LSB and all other outputs take place at every LSB until full scale. At the top of the ladder, REF(+) should not be more positive than the supply; and on the bottom of the ladder, REF(−) should not be more negative than ground.
- **Successive approximation register (SAR).** The SAR goes through eight iterations before it approximates the input voltage. Therefore, an n-iteration is required for any n-bit converter, regardless of the type of SAR. This register will be reset on the positive edge of the start conversion pulse (pin #6) and the conversion begins on the falling edge of this pulse (see the device timing diagram in Figure 6.14 for further detail).
- **Comparator.** The comparator is the most relied-upon element in the system. The accuracy of the whole operation of the system depends upon it. A chopper-stabilizer comparator has been chosen to do this work. The chopper will first convert the DC input signal to an AC signal. The signal is passed through an AC amplifier where its DC level will be stored. This processing method makes the converter totally immune to temperature, input offset errors, and long-term drift errors.

6.6 IIH Test (Logical "1" Input Current)

Definition

The sum of current passing through input protection diodes for any input condition that sets the output to logic high state.

Purpose

To verify the β value of the input transistor and its reverse emitter-to-collector operation.

Procedure

1. Force specified V_{CC} to the Power pin (pin #11), Ref pin (pin #12), and GND to 0 V.
2. Force the specified voltage at Start pin (pin #6).
3. Leave EOC (pin #7) open.
4. All other pins at 0 V.
5. Measure current at Start pin (pin #6).

6. Test the measured current against the specified maximum and minimum limits.
7. Force the specified voltage at any pin where the input current is to be measured.

Note. Make sure the EN pin (pin #9) is set low before the application of any signal to the output pin.

6.7 IIL Test (Logical "0" Output Current)

Definition

Sum of current passing through the input protection diodes for all input conditions that set the output to logic low state.

Purpose

Like the IIH test, this verifies the current passing through the input diodes and determines the existence of the base resistor.

Procedure

1. Force V_{CC} by the specified voltage.
2. Force 0 V to all the input pins that are to be tested. Maintain specified voltage at other input pins.
3. Leave EOC pin (pin #7) open during this test.
4. Measure current at any desired input pin.
5. Test the measured current.

Note. To avoid damage to the device, make sure pins 8, 14, 15 and 17 through 21 are left open.

6.8 IOZL Test (Three-state Low Level Output Leakage)

Definition

The measurement of the output current when the device is in high-impedance state.

Procedure

1. Force V_{CC} pin by the specified voltage and GND to 0 V.
2. Force specified voltage to EN (pin #9) to set the device to high-impedance mode.
3. Force 0 V to all the output pins that are to be measured for current.
4. Force specified voltage at all other output pins.
5. Run clock at the specified rate.
6. Force Start (pin #6) by applying signal with the defined amplitude.
7. Measure current at the specified output pins.

Note. The application of a signal at Start pin causes the device to start conversion. The conversion will be opposite to that in the IOZH test (the output tries to high).

6.9 IOZH Test (Three-state High Level Output Leakage)

Definition

The measurement of the output current at logic high state while the device is in high-impedance state.

Procedure

1. Force V_{CC} to the power and 0 V to GND pin.
2. Force Start (pin #6), Enable (pin #9) by the specified voltage to set the device to high-impedance mode.
3. Run Clock (pin #10) at the specified pulse rate.
4. Set all other pins to 0 V.
5. Apply a pulse with specified amplitude to the Start pin.

Note. The above setup enables the device to start conversion with low level output.

6.10 VOH Test (Logic "1" Output Voltage)

Definition

Minimum high level voltage at the device output for a given load at minimum V_{CC}.

Purpose

To verify the amount of load that the device output can handle (fan-out capability).

Procedure

A defined current sinking load must be connected to the output under measurement

1. Force V_{CC} pin with the proper operating voltage and 0 V to GND pin.
2. Apply input sequences to the input that sets the output to high logic state.
3. To enable the device to perform the code conversion, the following conditions must be set (refer to the device timing diagram):

   A. Clock should run by the defined frequency.
   B. Start pin (pin #6) should receive the defined pulse for beginning conversion.
   C. The Enable pin (pin #9) should be disabled.

4. Measure voltage at the desired output pins. Unused outputs should be left open.
5. Inputs must run by the application of the defined input voltage.
6. Test the measured voltage.

6.11 VOL Test (Logic "0" Output Voltage)

Definition

The maximum low level voltage at the device output for a defined load at minimum V_{CC}.

Purpose

To verify the fan-out drive ability of the device when the output is at logic low state.

Procedure

A defined current-sourcing load must be connected to the output under measurement.

1. Force V_{CC} to the power pin and 0 V to GND.
2. Set inputs to 0 V so the outputs stay low.
3. The condition of Clock, Enable and Start pins should be left as in VOH test.
4. Measure voltage at desired output pin(s). Unused outputs should be left open.
5. Test the measured voltage against the defined limit.

6.12 I_{CC}/I_{DD} Test (Power Supply Current)

Definition

Current passing through V_{CC} pin when the output is set at various logic states.

Purpose

Every output state draws different current through the device power pin. The current is measured under any given output logic state.

Procedure

No load is required.

1. Force the specified bias V_{CC} to the power pin and 0 V to GND.
2. Set all inputs to 0 V.
3. Run clock at the defined rate.
4. Force the defined voltage to Enable (pin #12).

5. Set the output(s) to open.
6. Measure current at the V_{CC} pin. The measured current must be within the minimum and maximum specified limits.

6.13 $I_{CC}+R$ Test (Combined $I_{CC}+R$ Current)

Definition

The total current passing through the power pin by the combination of power and reference voltages.

Purpose

To verify the amount of I_{CC}+resistor ladder current when the device is set to different output logic conditions.

Procedure

1. Apply V_{CC}+Ref voltage to the device power pin and 0 V to the GND pin.
2. Run clock at the defined frequency.
3. The rest of the set up condition is the same as for the I_{CC} test.
4. Measure current at the Power pin (#11). This current is the combination of the I_{CC} and the current across the resistor ladder.
5. Test the measured voltage against the specified limit.

6.14 $I_{IN(1)}$ Test (Analog High on Channel Input Current) while Clock is On

Definition

The measurement of current going through analog at high level input while the clock is running.

Procedure

1. Force V_{CC} by specified biasing voltage and 0 V to GND pins.
2. Run clock at the defined rate.
3. Force the specified input voltage only at the inputs that are going to be measured; set them to 0 V otherwise.
4. Apply the defined voltage at ALE (pin #22). Steps 3 and 4 enables the address decoder to receive the addresses.
5. Set all outputs to 0 V.
6. Measure current at the desired input pins. The measured current must be within the specified maximum and minimum.

6.15 $I_{IN(0)}$ Test (Analog Low on Channel Input Current) while Clock is On

Definition

The measurement of current going through analog inputs at low state when the clock is running.

Procedure

1. Force the specified biasing voltage to V_{CC} and 0 V to GND pins.
2. The rest of the setup condition is the same as the $I_{IN(1)}$ test except for 0 V at those inputs that are going to be measured; otherwise force the specified voltage to them.
3. The measured current, which is directed opposite to that in $I_{IN(1)}$, must be within the specified limits.

6.16 $I_{IN(1)}$ Test (Analog High on Channel Input Current) while Clock is Off

Definition

The measurement of current going through the analog input channel at high level while the clock is off.

Procedure

1. Force the specified biasing voltage to V_{CC} pin and GND to 0 V.
2. Apply no signal to the Clock pin.
3. The rest of the setup condition is exactly the same as the $I_{IN(1)}$ test described earlier. The measured current should agree with the predefined limits.

6.17 $I_{IN(0)}$ Test (Analog Low on Channel Input Current) while Clock is Off

Definition

The measurement of current going through the analog input channel at low state while the clock is off.

Procedure

1. Apply defined bias voltage to V_{CC} pin and set GND to 0 V.
2. Set clock to Off.
3. The rest of the test conditions are the same as the $I_{IN(0)}$ test described earlier.
4. Measure current at the desired input pin. The measured current should be within the defined limits.

6.18 Functional Test

Definition

To verify the device functional operation outlined by the manufacturer. The input stimulus and the predefined output response are determined by the logical function of the device based on the truth table or timing diagram.

Procedure

Since the accuracy, linearity, missing codes, and other functionality aspects of this device can be verified by this test, it is advisable to test the device for all possible patterns. There are eight inputs to the device where the input analogs are to be converted to their predefined binary values. The eight combinations of binary numbers create 256 different situations, in which every combination represents a defined voltage, and a 12-bit converter has 4095 codes. Adverse factors such as the D/A settling time, line capacitance, and comparators may cause nonlinearity in device testing.

To understand this test, it is necesssary to know the details of operation and the definitions of the parameters involved, as explained in the early pages of this chapter.

The AD809 is a unipolar A/D converter with voltage conversion range between 0 V and 5 V, where bits are changed every 19.53 mV:

$$\frac{5\,\text{V}}{256} = 19.53\,\text{mV}$$

Pin #21 is the most significant and pin #17 is the least significant output bit of the converter.

1. Set V_{CC} pins to the specified biasing voltage and GND pin to 0 V.
2. Set the defined pulse to Start pin (pin #6).
3. Run clock by the defined pulse rate and apply the required voltage at ENout (pin #9)

   A. Connect clock pulse to ALE (pin #22). This will enable the device to convert data continuously.
   B. Tie Start (pin #6) and EOC (pin #7) together.

4. Force 0 V to −REF (pin #16) and specified voltage to +REF (pin #12).
5. Apply the defined analog voltage (with respect to the device truth table, from 0 V to full scale and reverse) to the input pins. Each sequence of these input voltages is within the range of two expected subsequent binary bits at the output (see Note 3 below).
6. The resulting binary train must be within the maximum and minimum expected binary digits.

Note 1. The +REF voltage will change as the ambient temperature changes. The appropriate voltage condition is given by the manufacturer (usually room temperature).

Note 2. The sets of relays (mounted on the test load board) that give the choices of address line select must be controlled by software to correspond with analog channels.
Note 3. The input of A/D is an analog voltage, which must be converted to equivalent binary digits or equivalent decimal numbers. There is no way that binary "0" and "1" can be read from the output pins. The best technique that is common for this particular test is to connect the output of the A/D to the input of a permanently mounted D/A on the load board. The output of this device would then be a readable voltage corresponding to the analog input applied at the DUT inputs. The D/A used for this purpose must operate superbly. The operating voltage applied to this device as well as the analog input voltage applied to the DUT input must be very accurate. The total integrity of the D/A and the voltage regulators mounted on the load board must be verified before running any test.

6.19 T_{WALE} Test (ALE Pulse Width)

Definition

Minimum ALE is the narrowest pulse on the ALE pin that is guaranteed to latch in a new MUX address.

Procedure

Refer to the device truth table on the device data sheets (Figure 6.14).

1. Set the defined bias voltage to V_{CC} pin and set GND to 0 V.
2. Run clock at the defined frequency and force specified +REF or −REF to the related pins.
3. The specified *RC* timing network must be connected to pins 8, 14, 15, 17 through 21.
4. Measure the width of the pulse. The specification of the *RC* timing network is also given in the manufacturer's data book. The pulse width measurement in Chapter 3 may be helpful in understanding this test. The measured pulse width must agree with the predefined limit.
5. Measure the pulse width at each of the above pins simultaneously (from 50 per cent of the rising edge to 50 percent of the falling edge of the pulse or vice versa at the Start pin). It is apparent that the only timing network on the load board must be decoded to all the pins mentioned in Step 3 by the aid of relays, where their operation is controlled by the software program.

6.20 RES Test (Resolution Test)

Definition

The resolution of an *n*-bit converter sets the size of the least significant bit that the converter is able to quantize with respect to the converter's full-scale range. The resolution defines the number of "pieces" that the full-scale range can be broken into. The higher the resolution, the smaller the size of the LSB. For a binary type converter with 8-bit resolution, the size of the LSB is (full-scale range)/2^8. For a BCD type converter with three half-digit resolution, the size of the LSB is (full-scale range)/1999*2 counts.

Purpose

The purpose of the resolution test is to insure that the converter is able to generate usable information about incremental changes in the input voltage, and that there are no missing codes. Often a converter's internal noise limits its ability to quantize small incremental changes in input voltage and will be specified with an "effective number of bits" (ENOB) resolution.

Procedure

1. Force V_{CC} volts to the power pin and 0 V to the GND pin.
2. Force specified +REF and −REF voltage to the related pins.
3. Force OE (pin #9), ALE (pin #22) and Start (pin #6) by the predefined voltage.
4. Force 0 V to ADDA through ADDC pins.
5. Run clock at the specified frequency rate.
6. Enable the three-state condition.
7. Apply quantum voltages to any of the IN0 through IN7 inputs. The applied voltages must be accurate and noiseless within the accuracy of LSB.
8. Test the linearity of the device by the defined number of binary digits at the outputs of the device.

6.21 Nonlinearity Test

Definition

A converter's nonlinearity is a measure of the deviation of the device transfer curve from a straight line drawn through the zero code to full-scale code transition on a digital output code vs. analog input scale. An ideal converter will increment one digital code for each 1 LSB increase in analog input voltage.

Purpose

A converter's nonlinearity is an untrimmable error and defines how accurately the converter can measure (quantize) any given input voltage. The converter's nonlinearity is also related to the expected signal-to-distortion ratio, as nonlinearities will cause harmonic distortion.

Procedure

1. Apply V_{CC} volts to the power pin and 0 V to the GND pin.
2. The rest of the test setup is the same as for the resolution test. The linearity of the device as defined by the data book can be verified by the number of binary outputs. The simplest way of testing the device is to verify the linearity from 0 V to full scale. However, this is not a sure strategy for converter testing. Nonlinearity and missing codes may occur between 0 V and midway along the line or from there to the

full-scale point. Although time-consuming, the best method is to check for as many points as possible to assure the integrity of the intended function. The summation codes method can drastically reduce test time. In this method code bits are usually divided into two groups, those having positive bit errors and those having negative bit errors. The device is tested for linearity using these two groups simultaneously.

3. The linearity of an ADC is sometimes verified by the application of three different voltages that represent three distinct points along the linearity line from 0 V to full scale. Zero volt and the full scale would be two of these candidates, the middle of the line would probably be the next best choice. The output voltages corresponding to these three points must be located on the linearity line as shown in Figure 6.6. The unit of measurement is related to $\pm\frac{1}{2}$ LSB.

6.22 RL Test (Ladder Resistance Test)

Definition

The ladder resistance is the resistance in ohms as measured between the reference (+) and (−) pins.

Purpose

The purpose of this test is to define the amount of current an external reference device must provide to drive the converter's internal resistor ladder.

Procedure

1. Force the power pin and ground pin by V_{CC} and 0 V, respectively.
2. Enable the three-state condition by the application of the defined voltage at the appropriate pin.
3. Apply defined voltage to the Start pin.
4. Enable the MUX input by the application of the predefined voltage at the proper pin.
5. Run clock by the defined rate.
6. Measure the resistance between −REF and +REF (pins #12 and #16). The measured value must be within the manufacturer's suggested limits.

6.23 T_{WS} Test (Minimum Start Pulse Width)

Definition

Minimum start pulse width is the narrowest pulse on the Start pin that will insure that the converter initiates a conversion.

Purpose

If the given pulse is too narrow, the converter may not respond and no new conversion will occur.

Procedure

1. Apply the required V_{CC} to power pin and 0 V to the GND pin.
2. Set the required +REF and −REF voltages to the related pins.
3. Run clock at the defined frequency.
4. Specified RC timing network must be connected to pins 8, 14, 15, 17 through 21. The specified resistor and capacitor must be accurate (see pulse width test in Chapter 3, Section 3.18).
5. The setup for the width test requires the implementation of the device timing diagram shown in the manufacturer's data sheets.
6. Measure pulse width from 50 percent of the rising edge to 50 percent of the falling edge of the pulse at the Start pin.
7. The measured value should agree with the manufacturer's limit.

6.24 T_S Test (Minimum Address Setup Time)

Definition

Minimum amount of time a valid address must be on the MUX address lines before the rising edge of the ALE pulse occurs to insure that the correct address is latched in.

Procedure

The same setup as for the T_{WS} test is needed. The measurement is now taken from 50 percent of the rising edge pulse at the ALE pin to 50 percent of rising or falling edge of each address pin simultaneously. The measured time should satisfy the predefined requirement.

6.25 T_H Test (Minimum Address Hold Time)

Definition

The minimum amount of time a valid address must be held stable after the rising edge of ALE. If the MUX address is not stable during the period, an incorrect address may be latched.

Procedure

The setup that was used for the T_{WS} test is also used for this test. The measurement is taken from 50 percent of the rising edge of the pulse at the ALE pin to 50 percent of the rising edge of the pulse at the Start pin. The measured time should satisfy the minimum predefined requirement.

6.26 T_D Test (Analog MUX Delay Time from ALE)

Definition

The delay time from the rising edge of ALE to the MUX actually completing switching to a new channel (input of the comparator settles to less than $\frac{1}{2}$ LSB of new value). New Channel conversion data must be discarded until this time has passed.

Procedure

The T_D is longer than the previous timing measurements and therefore the *RC* timing network must be replaced with the one that has the longer time span as defined in the manufacturer's data book. This measurement verifies that the MUX output goes through one transition and that transition occurs within 50 percent of the rising edge of the pulse at ALE to 50 percent of the falling edge of the pulse at the Start pin minus the time for $\frac{1}{2}$ LSB. The transition may be from low to high or vice versa. The minimum delay must be within the given limits.

6.27 T_{H1}, T_{H0} Test (OE Control to Q Logic State)

Definition

OE to Q logic state is the time required between the rising edge of OE and the converter's digital outputs achieving a valid logic state (three-state enabled).

Procedure

The *RC* timing circuit must be modified according to the manufacturer's data sheet. The setup used for this test the same as that used for the T_{WS} test.

1. For T_{H0} the measurement is made from 50 percent of the rising edge of the pulse at the OE pin to 50 percent of the falling edge of the next pulse at any output pin.
2. For T_{H1} the measurement is made from 50 percent of the rising edge of the pulse at the OE pin to 50 percent of the rising edge of the next pulse at any output pin.

The measured time must be within the expected limits.

6.28 T_{H1}, T_{H0} Test (OE Control to High-*Z*)

Definition

The amount of time required for the converter's digital output pins to go to a high-impedance state after the falling edge of OE (on to three-state). In a bus type system, this is the amount of time the converter will take to relinquish the bus to another device or operation.

Procedure

The previous test setup is required for this test with the exception that the three-state condition must be maintained. For T_{H0}, the measurement is done at 50 percent of the falling edge of the pulse at the OE pin to 50 percent of the falling edge of the pulse at any of the outputs. The measured time must satisfy the requirement.

6.29 T_C Test (Conversion Time)

Definition

Conversion time is the time interval required by the converter to complete one conversion (quantization) of the input signal. For an SAR type converter, this time is directly proportional to the clock frequency, as a specified number of clock cycles is required to complete one conversion.

Procedure

The clock frequency must be maintained at the defined rate. No three-state condition is required and the test setup is the same as for the T_{WS} test. Since the conversion time is related to Start and EOC, the measurements are taken from these two pins. This time is measured from 50 percent of the falling edge of the pulse at the Start pin to 50 percent of the rising edge of the pulse at the EOC pin.

Note. The data register output should be updated just one clock pulse before the rising edge of the pulse at the EOC pin.

6.30 F_C Test (Clock Frequency)

Definition

Clock frequency is the minimum and the maximum operating frequency of the clock applied to the converter. Exceeding these limits may cause the converter to function incorrectly or no longer to meet resolution and accuracy specifications.

Procedure

The same test setup as used for the T_{WS} test is required here. This test verifies that the frequency of the clock is maintained as intended. The test can be performed by measuring numbers of the clock period and then converting to frequency. The limit is expressed in hertz or kilohertz.

References

1. Ralph Joice, *Operational Amplifier*, National Semiconductor Corp.
2. *Burr-Brown IC Data Book, DAC1200KP-V*, Vol. 33
3. DAC-02/DAC-03/DAC-05, Precision Monolithic, Inc. Data Book, 8/87, Rev. B, 1987.
4. *Linear Integrated Circuits Data Book, Types ADC0816, ADC0817*, Data Acquisition Systems, Texas Instruments, Inc., 1989.
5. *Samsung Electronics Data Book, KAD0808/KAD0809*, Samsung.
6. *ADC0808/ADC0809 Data Book*, Texas Instruments, 1989.
7. Jim Sherwin, *Specifying A/D and D/A Converters, AN-156*, National Semiconductor Corp, 1989.
8. Larry Wakeman, *Using ADC0808/ADC0809 8-bit mp, AN #247*, National Semiconductor Corp., 1989.
9. *Signetics Analog Manual, Application Specifications*, Signetic Corp., 1976.
10. Eugene R. Hnatek, *A User's Handbook of D/A and A/D Converters*, Wiley, 1988.

11. *Analog-Digital Conversion Handbook*, Analog Devices, Inc., 1986.
12. Bernard Loriferne, *Analog-Digital and Digital-Analog Conversion*, Wiley, 1982.
13. David J. DeFatta, Joseph G. Lucas and William S. Hodgkiss, *Digital Signal Processing: A System Design Approach*, Wiley, 1980.
14. Data Acquisition and Conversion, Databook Vol. 2, *Linear Circuits*, Texas Instruments, Inc., 1989.

CHAPTER 7

Digital Signal Processing

Introduction

This chapter describes digital signal processing (DSP) in some detail. The absence of unnecessary complex mathematical formulas renders the topic more comprehensible, than might be anticipated. The subjects covered here are extremely important for the follow-up discussion in Chapter 8, which deals with the CODEC communication device. Some twenty technical references have been reviewed to derive this chapter in a fashion appropriate for the use of semiconductor test engineers.

Topics Digital signal processing, introduction.

7.1. to 7.3. Function sample and hold modules, audio digitizers, introduction
7.4. Function generator, including amplitude and phase spectra, cycle of sine waves
7.5. Description of Fourier series and transforms along with schematics and a few simple formulas to identify the time or frequency domain in each spectrum
7.6. Signal conversion and the vital role of the Nyquist theorem in choosing the sampling rate; procedure for coherent and noncoherent sampling and the problems involved
7.7. Sampling rate, sampling process, and companding
7.8. Errors in use of Fourier series or transform (aliasing, leakage or picket-fence effect)

Devices that convert analog signals to digital (A/D)) and digital signals to analog (D/A) are the backbone of digital signal processing (DSP). DSP in automated test equipment (ATE) is a microprocessor-controlled operation that processes signals at high rates. DSP is broadly used in telecommunications, especially integrated service digital networks (ISDN), filtering and speech recognition, and comparison and medical imaging such as magnetic resonance imaging (MRI).

Fourier analysis (Fourier series and transform) is the fundamental tool used today to characterize linear or mixed-signal microchips. A signal sampled by this method can subsequently be analyzed for a variety of complex parameters such as signal-to-noise ratio (SNR), total harmonic distortion (THD), phase and gain discrepancies in the frequency domain. To achieve this, the discrete Fourier transform (DFT) and the fast Fourier transform (FFT) are employed; both produce the same results with minor differences in speed. The discrete Fourier transform and the fast Fourier transform are both widely used to measure the parameters mentioned above from a train of data.

The Fourier series and Fourier transform are mathematical tools employed to compute and analyze complex signals. Modern, mixed-signal, ATE can rapidly and accurately perform various tests on analog signals capable of generating a variety of waveforms.

- The Fourier transform is used to analyze a signal sample in the frequency or the time domain. It can be either forward or reverse.
- The Fourier spectrum is a series of discrete components such as a fundamental frequency and its integer multiples.

The highest frequency the FFT or DFT can be used to measure is half the sampling rate frequency.

The relationship between the sample number and the frequency components in the Fourier series is

$$f_s = \frac{N}{2}$$

where N is the number of samples.

The spectrum is discrete for periodic input signals and continuous for nonperiodic signals. In the structure of most mixed-signal ATE, there exist a number of subsystems; each of these subsystems is responsible for performing a part of the signal processing procedure as dictated by the mathematical theory. For clarity, we first identify these components in a brief introduction; as we proceed, the reader will become more familiar with the application of Fourier series or transforms using these hardware modules.

7.1 Anti-aliasing Filter

Aliasing is the occurrence of unwanted frequencies at the output of the A/D or PCM (pulse-code modulation) system that did not exist at the input. Foldover of higher frequencies is the cause. A low-pass filter will suppress higher-frequency components of an incoming signal prior to the sampling stage if the spectrum and associated signal noise are above the acceptable frequency and are nonzero (more details given in Section 7.7).

7.2 Sample-and-Hold Module

The sample-and-hold module is made up of a capacitor that is activated by the sample rate pulse and is charged to a level equal to the voltage of the input sample. The hold sample is then moved (by the capacitor discharge) to the next stage of the process and the capacitor is recharged by the next segment of the input signal. The purpose of this module is to reduce conversion time for the conversion of varying incoming signals. This module will reduce the time error (called uncertainty error) by the ratio of the A/D conversion operation time to the time required by the sample-and-hold module (more details given in Section 7.6.2).

The functions of A/D and D/A converters were explained in detail in Chapter 6, which is devoted to them.

7.3 Audio Digitizer

An audio digitizer is a module with a high-speed, high-resolution A/D converter. It digitizes an analog input signal and stores it in accessible memory as a sequence of digits. These digits represent the amplitudes of the voltage captured by the sample-and-hold circuit at designated time intervals.

The following gives very brief overview of a DSP module in a modern mixed-signal ATE. We begin with the fundamental aspects of a signal and follow it through the process of becoming a measurable signal sample to a digital computer. The process requires some mathematical expressions familiar to any engineer.

7.4 Function Generator and Signal Properties

In modern ATE the function generator is a programmable D/A converter that is capable of generating waveforms from digital data. The functions of the D/A and A/D converters are controlled by the clock pulses present in both modules.

The sine wave, a basic phenomenon of the universe, can be defined by its frequency or its amplitude. The frequency, which is the number of cycles per second is expressed in hertz (1 Hz = 1 cycle/s), and amplitude is measured in volts as voltage peak or peak-to-peak. On this basis the signal can be described in either the frequency domain or the time domain. In the first case a signal consists of a sinusoidal wave of specific frequency.

Figure 7.1 shows a full cycle of a sine wave as instantaneous magnitude v versus angle θ, where θ varies from 0 to 2π radians (0 to 360 degrees).

$$v(\theta) = V_m \sin\theta$$

V_m, a constant, is the maximum amplitude.

The time component can be expressed trigonometrically using the relation

$$\theta = \omega t$$

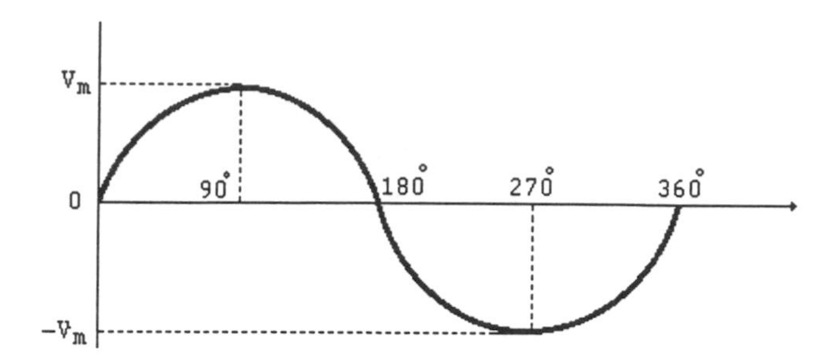

Figure 7.1 A full cycle of a sine wave: $v = V_m \sin\theta$.

where ω is the angular frequency (radian/s). The period T is the time taken for one full cycle of 2π radians: time variation from 0 to T corresponds to angle variation from 0 to 2π radians. Thus frequency f (Hz), period T (seconds) and angular frequency are related as

$$T = \frac{1}{f} = \frac{2\pi}{\omega} \qquad \text{i.e.,} \qquad \omega = 2\pi f \quad \text{or} \quad \omega = \frac{2\pi}{T}$$

These relations allow the alternative descriptions of a sine wave:

$$v(\theta) = V_m \sin\theta$$

$$v(t) = V_m \sin(\omega t)$$

$$v(t) = V_m \sin(2\pi f\, t)$$

$$v(t) = V_m \sin\left(\frac{2\pi}{T} t\right)$$

A change in angle θ by 90 degrees in either direction results in the sine wave becoming a cosine wave. Addition of a positive phase angle ϕ results in a wave that is said to be leading the original wave by ϕ. The general expression of a sinusoidal wave is therefore

$$v = V_m \sin(\omega t + \phi) \tag{1}$$

Sinusoidal waves with the same frequency can be added together trigonometrically.

In conversion or transmission of a train of sine waves from one place to another we deal with the frequency, amplitude, and phase angle, which comprise a complete identification of a sinusoidal wave. In the following, f_t is the carrier frequency and θ is the carrier phase angle, $x(t)$ is the instantaneous magnitude and A is the carrier amplitude:

$$x(t) = A\cos(2\pi f_t\, t + \theta) \tag{2}$$

This relation is represented as spectra in Figure 7.2. It can be represented as an

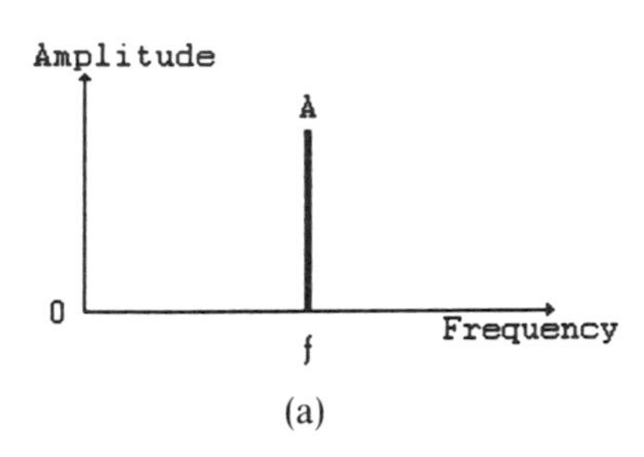

(a)

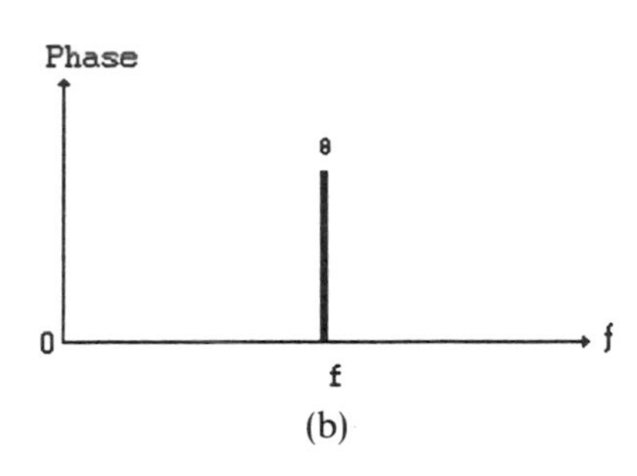

(b)

Figure 7.2 (a) Amplitude and (b) phase vs. frequency representation of a signal.

amplitude spectrum (amplitude vs. frequency) or as a phase spectrum (phase vs. frequency). The carrier amplitude is changed in amplitude modulation; f_t is changed in frequency modulation; in phase modulation the phase angle θ is changed. Regardless of which approach is chosen, the result is a new frequency band of the processed signal centered around the original frequency.

7.5 Fourier Series and Fourier Transforms

Any train of pulses (including human speech) can be decomposed into series of sinusoidal pulses that vary in frequency or amplitude. The third component of a sine wave is the phase angle, which has to be considered when two or more different sine waves with same frequencies are compared to one another. The phase angle determines whether one wave is leading or lagging another.

In the frequency modulation method, a signal is analyzed in terms of its frequency. *Fourier series* and *Fourier transforms* are the best tools available today to describe a signal mathematically. The Fourier transform is used to determine the frequencies of nonperiodic waveforms with finite energy in each period. For signals having periodic changes, containing finite energy in a finite interval, the Fourier series is used.

Thus, using the Fourier series, any periodic function can be written as an infinite series of harmonically related, sinusoidal terms. Consider the waves shown in Figure 7.3. Although they are not sinusoidal, these waves are periodic with period T and all have the same amplitude A. The Fourier series enables us to write these periodic functions as a series of one-line mathematical expressions containing sines and cosines:

$$X(t) = a_0 + (a_1 \sin \omega_0 t + b_1 \sin \omega_0 t) + (a_2 \cos 2\omega_0 t + b_2 \sin 2\omega_0 t) + \cdots \quad (3)$$

This expression can be simplified using summation notation:

$$X(t) = a_0 + \sum_{n=1}^{\infty} (a_n \cos n\omega_0 t + b_n \sin n\omega_0 t)$$

where a and b are series coefficients in the above relation. The later terms of the

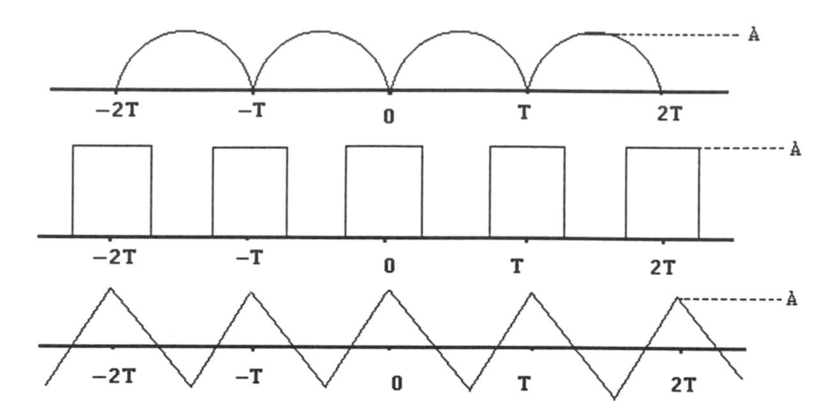

Figure 7.3 Different waves with similar period.

series are truncated and the amplitude of the harmonics decreases in the progression. The power of the Fourier series can be described as *infinite*. In the series above,

$$\omega_0 = 2\pi f_0 = \frac{2\pi}{T}$$

Fundamental frequency is the term used for the wave of period T. The fundamental frequency of the first term in Equation (3) is 0. The second, third, and fourth components, are called *harmonics*, where $2f_0$, $3f_0$, and $4f_0$ are the second, third, and fourth harmonics, respectively. The amplitude of each harmonic is lower than that of the preceding one. On this basis, the higher or the upper harmonics will be filtered out. The DSP internal filtering stage plays a major role in managing this process. This is achieved by design specification, and it is not hard to appreciate that the accuracy of the capacitor in this stage is extremely critical.

As stated, these harmonics decrease in amplitude as the series progresses. The coefficients $a_0, a_1, \ldots b_0, b_1, \ldots$ are determined by integration. Coefficient a_0 is the mean value of the $X(t)$ function within the rate that the signal is expanded. To determine the average value of the signal in a_0 (the first term), proceed as follows:

$$a_0 = \frac{1}{T} \int_{-T/2}^{T/2} X(t) dt$$

To attain the complete description of the periodic function with the aid of Fourier series, it is required to take the integration of all the components of the $X(t)$ function in the same manner as applied to the a_0 term. For instance, the mathematical process for the a_n and b_n coefficients in $X(t)$ is outlined as:

$$a_n = \frac{2}{T} \int_{-T/2}^{T/2} X(t) \cos n\omega_0 t \, dt, \qquad b_n = \frac{2}{T} \int_{-T/2}^{T/2} X(t) \sin n\omega_0 t \, dt$$

The component that possesses the lowest frequency (close to zero) is known as the fundamental frequency and is followed by the rest of its components (a_n, b_n, ..., in our case), called harmonics. To verify the description of a signal in the frequency domain, it is required to know the amplitude, frequency axis, and phase angle of all the harmonics. The integration of the second to the nth terms determines the amplitude of different harmonics at different frequencies. Fortunately, in the modern ATEs these operations are performed by electronic modules that are

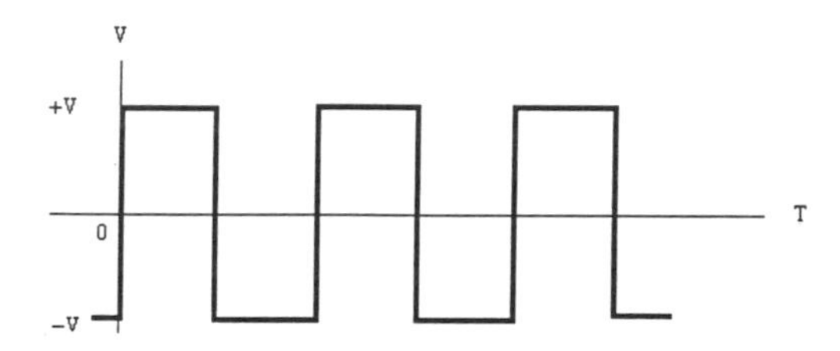

Figure 7.4 A square wave with amplitude of $\pm V$.

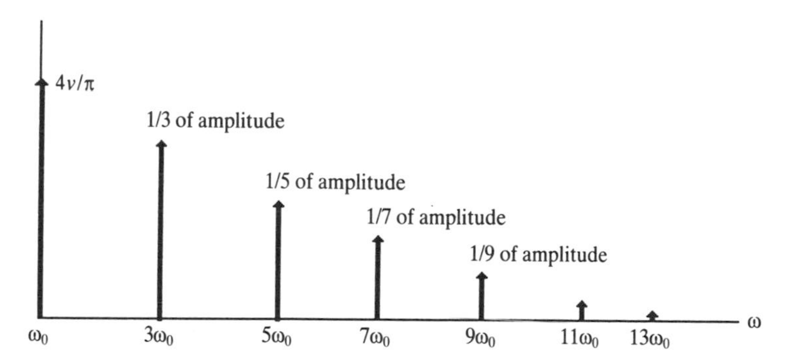

Figure 7.5 Spectrum of Figure 7.4.

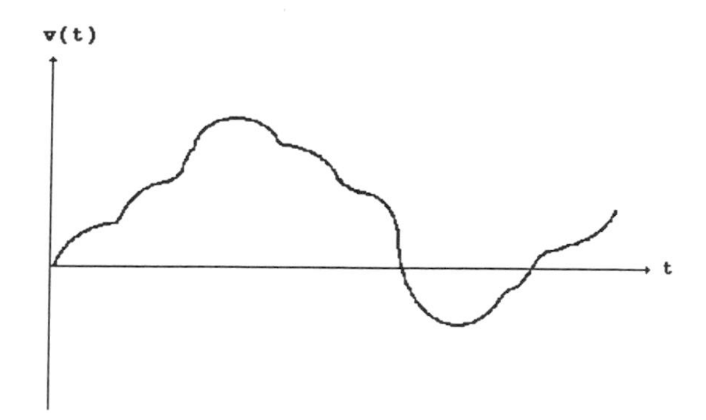

Figure 7.6 A nonperiodic function.

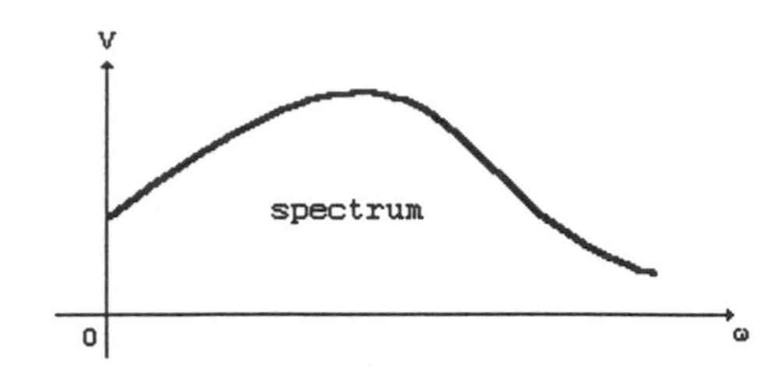

Figure 7.7 Spectrum of a nonperiodic function.

controlled by a high-speed microprocessor and the process is controlled by specialized software that uses firmware codes in the read-only memory (ROM).

Equation (3) contains the frequency spectrum of the pulses in the series. For instance, for the square wave with an amplitude voltage of $\pm v$ in Figure 7.4, the spectrum in Figure 7.5 can be obtained.

When the Fourier transform is applied to a nonperiodic function, it creates a continuous frequency function. Consider the signal wave shown in Figure 7.6 and its spectrum shown in Figure 7.7. The difference between Figure 7.7 and the spectrum for a periodic function, where each harmonic is represented as a discrete frequency, is that Figure 7.7 contains all the possible frequencies in the nonperiodic signal.

7.6 Signal Conversion

There are two different theories for communication signal conversion:

- Nyquist theorem
- Shannon channel capacity theorem

The Nyquist theorem, which will be discussed in the following sampling sections, deals with the maximum rate at which a train of pulses can be guided into a system communications channel without overlapping or interfering with one another. The Shannon theorem indicates the number of binary bits that can be transmitted per second to the communication channel without interference, called the *bit rate.*

It is probably not inappropriate to define the term baud rate, which we frequently hear when dealing with modems. The speed of the process by which the original signal is changed by the frequency, amplitude, or phase per unit of time is described by the *baud rate.*

The shape and amplitude values of an analog signal (as represented in Figure 7.6) can vary. A portion of an analog signal at specific times as a sequence of values is represented as in Figure 7.8b. These numbers represent digital codes. An audio digitizer creates these digitized codes, which are sent to an array processor module for processing into amplitude-versus-frequency data.

Sampling is the term used for transforming an analog signal to a discrete

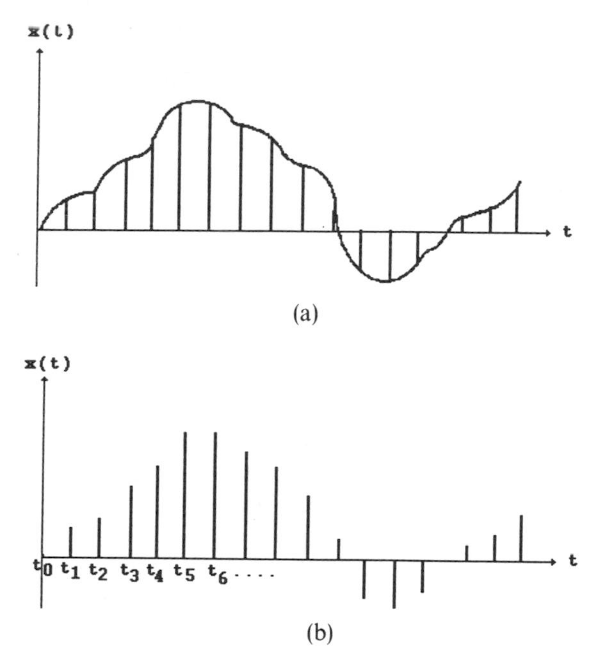

Figure 7.8 Digitization of an analog signal.

time signal by the analog to digital converter (ADC) in the audio digitizer section of the DSP system in modern mixed-signal ATE. The amplitude of the samples obtained is stored digitally as a set of numbers. *Modulation* is the process of shifting the frequency of a signal into a band suitable for communications equipment. Three types of modulation can be applied to a signal:

- Amplitude modulation
- Frequency modulation
- Phase modulation

Pulse code modulation (PCM), which is the most common method for digital pulse transmission, is the conversion of a signal from analog to digital by means of coding. It requires a large bandwidth for transmitting large numbers of pulses. PCM is largely used in satellite communications, industrial telemetry, and short-distance and heavy-capacity telephone circuits.

Pulse code modulation consists of the following steps:

1. Sampling of analog signal (discussed later in the next section)
2. Quantization of the sample signal (discussed in chapter 6)
3. Encoding the quantized signal into digital codes

The two different types of sampling in signal processing are

1. Coherent sampling
2. Noncoherent sampling

7.6.1 Coherent Sampling (Synchronous)

Coherence of a function is a property of the frequency domain and is represented by its signal-to-noise ratio. More frequent sampling is required for rapidly changing signals in order to adequately represent the signal digitally.

The sampling rate of a signal is determined by its total bandwidth. The sampling rate must be higher than the highest frequency in the bandwidth of the input signal in order for all the signal information to be sampled. If a signal has no frequency higher than the ω_0 fundamental frequency, then the signal can be taken at any interval where

$$T_N \leqslant \frac{\pi}{\omega_0}$$

The Nyquist theorem states that the sampling of this type of signal has a limit determined by $T_N = \pi/\omega_0$. This is called the *Nyquist interval*, and is expressed in seconds per sample. Since the telephone voice channel is band-limited to less than 4 kHz (the highest frequency in regular speech), then 8K samples/second, providing 7 or 8 bits per sample, will be adequate to guarantee the quality of the original signal. These bits can be transmitted at 64K bits/second. (For more information on this subject, the reader may refer to any book on communication.)

The frequency $\omega_N = 2\omega_0$ is called the *Nyquist frequency*, which is one-half of the input frequency. The conclusion is:

> In order not to lose any information about the input signal, the frequency of sampling rate must be at least twice the input frequency

When the arriving signal and the sampling rate are in phase with one another, they are coherent.

7.6.2 Noncoherent Sampling (Asynchronized)

There is no synchronization between the phase angle of the arriving waveforms and the sampling rate in the case of noncoherent sampling.

Consider the waveform in Figure 7.8a (where the spectrum rolls off from 200 Hz to 3500 Hz, audio frequency) being sampled at 9000 samples per second as shown in Figure 7.8b. A continuous time function, which follows the original waveform, is obtained if the original portion of the spectrum is saved and the rest is truncated by a reconstruction filter. In Figure 7.8b the instants in time when the amplitude of the analog signal captured (by the track-and-hold module) and measured are represented by $t_0, t_1, t_2, \ldots$.

The A/D conversion process generates a series of these samples, or numbers, that represent the amplitude and form of the original input signal. In the A/D process, the general rule is

The higher the frequency of the input signal, the smaller the sampling interval required.

The DSP capacitor obtains samples by charging and discharging, according to the system clock sampling rate assigned to it. The charged capacitor holds the energy contained in a sample signal when the switch is closed. The frequency of the digitizer interval clock regulates the length of time the capacitor can be charged and held. In Figure 7.8b, discrete charges of the amplitudes indicated are stored in the capacitor at clock intervals $t_0, t_1, \ldots$.

A waveform is sampled and digitized with very minor delays as new inputs continuously arrive for sampling. Distortion can occur if the inputs arrive at a rate approaching the sample rate. The distortion caused by this lack of coherence between the arriving waveforms and sampling rate is called *leakage* or the *picket-fence effect*, which will be discussed later in Section 7.8.2.

The amplitude of a sampled signal can take an infinite number of values, but because a digital pulse can only be represented by a discrete value, it is necessary to quantize the amplitude of a stored sample. *Quantization* is the process by which the continuous amplitude range of the sampled signal (analog wave) is converted into a series of binary values as close as possible to the actual amplitude of the arriving signal. The sample signal is then converted into binary codes. The resulting codes are not coherent, either in time or in amplitude. The next step is to transmit each 8-bit signal of each sample together with appropriate spacing between each. The result is a digital sequence equivalent to the original input analog signal that is understandable to a digital system. This will be clearer when we discuss the operation of the coder/decoder in Chapter 8. The main goal of the discussion at this point is to give the reader an idea without mathematical theory of the processes that an analog signal goes through until it becomes an equivalent digital representation that can be manipulated by a digital computer.

7.7 Sampling Rate

The periodic frequency spectrum of any nonperiodic pulse can be obtained by using Fourier transformation as shown in Figure 7.9.

The process illustrated results in a new spectrum whose period equals the sampling rate. The new spectrum is actually the original signal replicated along the frequency axis. The sampling rate used for the process to be effective in creating such a spectrum must be at least the minimum rate adequate for accuracy that is economic in implementation. Figure 7.10 depicts the sampling process.

To digitize speech, the sampling rate must be frequent enough to contain all of the speech characteristics, but not so frequent that it becomes overwhelmingly costly. For example, a rate of a few samples per second would not adequately sample speech, and 10^6 samples per second would be far too frequent to be practical. However, any signal has an optimum sampling rate that adequately captures the

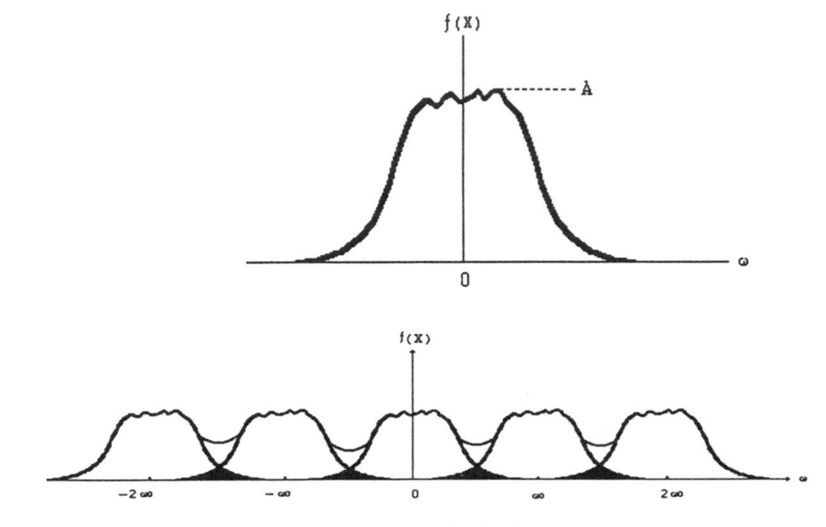

Figure 7.9 Schematic of a folded spectrum.

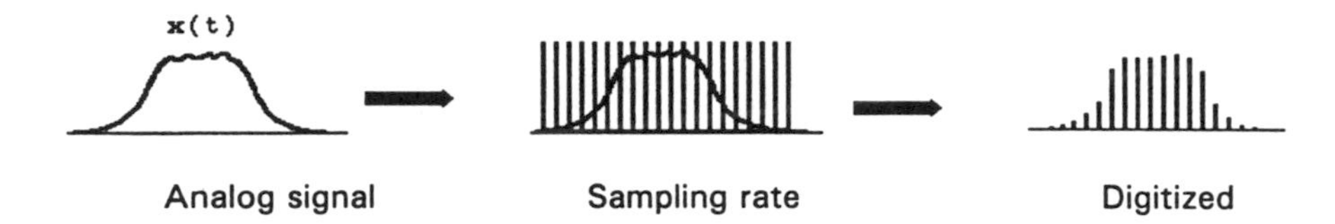

Figure 7.10 Relation between the signal and sampling rate.

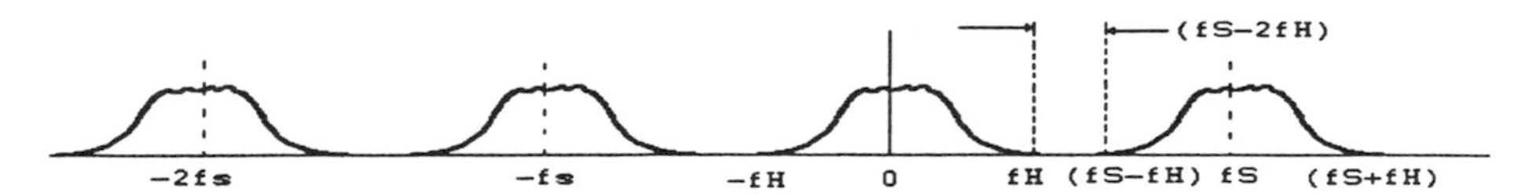

Figure 7.11 Sampling rate for digitizing speech.

detail of the spectrum. In Figure 7.11 the sampling rate f_s and the center frequency of the pulse are equal; the edge-to-edge spacing indicated is $f_s - 2f_H$.

The spectrum will contain all the required information as long as the sampling rate is $f_s \geqslant 2f_H$. A low-pass filtering operation is used to recover the information of the original signal. This operation may not be possible if $f_s < 2f_H$. A sampling rate less than $2f_H$ may cause the individual spectrum bands to overlap as shown in Figure 7.9. This overlap is called *aliasing*. The condition for $f_s \geqslant 2f_H$ is known as the *Nyquist rate*.

Digitization of sample signal by an internal DSP's A/D module is the *sampling process*. Compression and expansion of a signal, or *companding*, is a technique for improving the resolution of a sampled signal. The digitizer module has a compressor on the transmitter side and an expander on the receiver side. More details of the companding process will be given in the next chapter.

7.8 Errors

When using the Fourier series or Fourier transform to convert an analog signal to equivalent digital signal code, the computable form is subject to certain errors and adjustments that cause differences between the original waveforms and the computed form. Some errors the test engineer must be prepared to deal with are

- Aliasing
- Leakage or picket-fence effect

7.8.1 Aliasing

Unless a reference signal is bandlimited to half of the sampling rate (the Nyquist theorem) it cannot be sampled without distortion. The continuous transform time signal of $x(f_1)$ shown in Figure 7.12 is bandlimited to $0 < f < f_H$. If this signal is sampled at a rate of $f_s < f_H$, its transform $x_2(f)$ exhibits an overlap, an aliased spectrum as shown in Figure 7.13. The aliased signal is formed by the bits between the signal and the sampling rates.

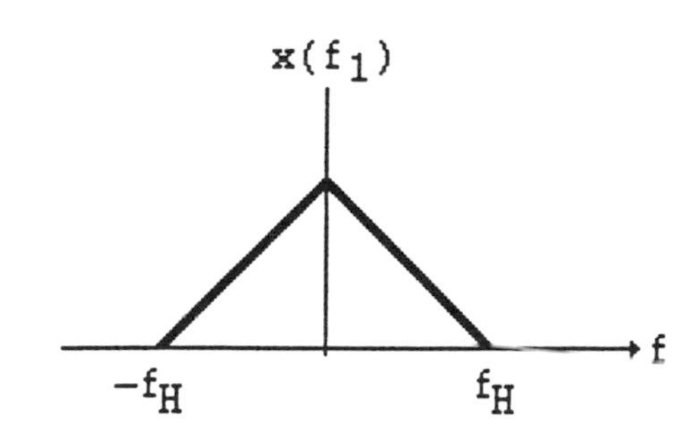

Figure 7.12 A sample signal.

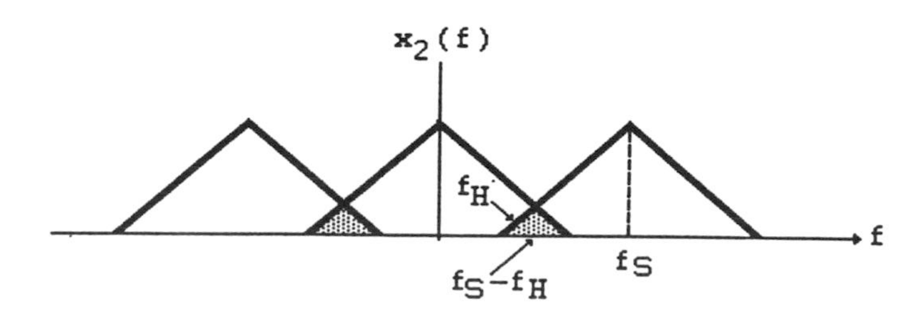

Figure 7.13 Aliased spectrum.

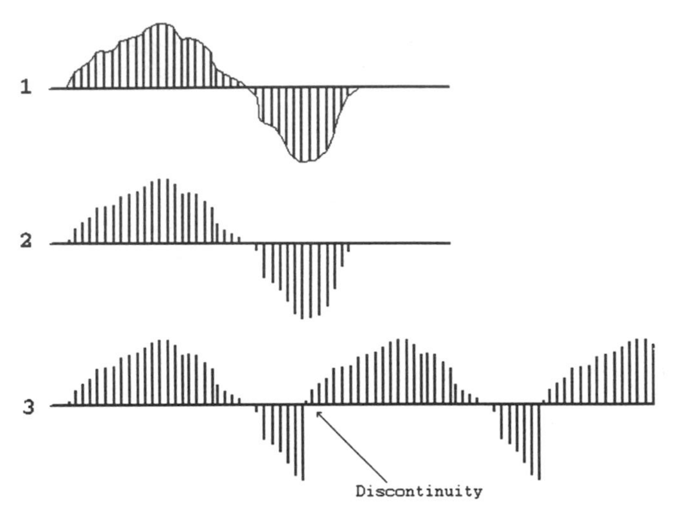

Figure 7.14 Leakage during sampling.

Other frequencies can sometimes be mistaken as the overlap frequency. In our example, the sampling rate and the signal frequency can be mistaken for the overlap frequency. A sampling rate large enough to avoid these being mistaken for other frequencies is the solution. The specification of the input signal must be fully understood before proceeding with the sampling rate determination.

In some cases it helps to filter the analog signal using a low-pass filter before sampling. Most mixed-signal ATEs are equipped with software-controlled filters.

7.8.2 Leakage or Picket-fence Effect

The DFT is sometimes unable to see the spectrum as a continuous function because the spectral repetition spacing is dependent on an integer multiple of the fundamental frequency. If this happens, a component peak point that occurs between two discrete transform lines may not be detected.

Varying the number of points in a time period is one way to prevent this error; this changes the location of the spectral lines without disturbing the shape of the original spectrum.

When analyzing a signal that is not periodic and not time-limited, leakage occurs because the digitized signal is longer than the input to the discrete Fourier transform (DFT). The extra length is truncated before the DFT can be applied and the DFT, being periodic, assumes that the truncated portion of the signal is a part of the same continuous sequence. The discontinuity that results at the end of the input signal is called *leakage* and is shown in Figure 7.14.

Leakage is best approached by choosing a suitable window function to shift the signal to the left by $N/2$ and minimize the error. N is the number of samples in the range

$$0 \leqslant F \leqslant F_s$$

where the frequency function is sampled, or

$$0 \leqslant t \leqslant t_p$$

where the time function is sampled.

References

1. N. B. Jones (ed.), *Digital Signal Processing*, IEEE Control Engineering Series 22, Peter Peregrinus, 1982.
2. Thomas Young, *Linear Systems and Digital Signal Processing*, Prentice-Hall, 1985.
3. Samuel D. Stearns, *Digital Signal Processing*, Hayden, 1985.
4. I. T. Young, J. Biemond, R. P. W. Duin, and J. J. Gerbrand, *Signal Processing III, Theories and Applications*, 3rd European Signal Processing Conference, North-Holland, 1984.
5. William D. Stanley, *Digital Signal Processing*, Reston, 1984.
6. William D. Stanley, Gary R. Dougherty, and Ray Dougherty, *Digital Signal Processing*, 2nd ed., Reston, 1984.
7. *Linear Circuits, Data Acquisition and Conversion*, Vol. 2, Texas Instruments, Inc., 1986.
8. Robert M. Gagliardi, *Introduction to Communication Engineering*, 2nd ed., Wiley, 1978.
9. C. Britton Rorabaugh, *Communication Formulas & Algorithms*, McGraw-Hill, 1990.
10. Bertil C. Lindberg, *Troubleshooting Communications Facilities*, Wiley, 1990.
11. Kamilo Feher/Engineers of Hewlett-Packard, *Telecommunications Measurements, Analysis, and Instrumentation*, Prentice-Hall, 1987.
12. Ronald L. Fante, *Signal Analysis and Estimation*, Wiley, 1988.

CHAPTER 8

CODEC (Coder/Decoder)

Introduction

The coder/decoder (CODEC) device is the backbone of telephone communications. The best known and most important parameters in CODEC testing are outlined in this chapter. These parameters include total harmonic distortion (THD) and signal-to-noise ratio (SNR), which require the DSP process as well as the gain error. A schematic of an on-bench test connection to test the above parameters in single-tone methodology along with the explanation of the circuit operation is presented in this chapter.

Topics Brief description of the structure of the device, and its functions.

8.1. Frame concept, including details of the CCITT and AT&T regulations
8.2. Companding rules in communication, as determined by the Consultative Committee for International Telegraph and Telephone (CCITT)—the A-87.6 law; and American Telephone and Telegraph (AT&T)—the μ-255 law; description of μ- and A-laws along with detailed schematics; differences between communication laws
8.3. C-messages and P-messages for noise, and the related graph
8.4. Unit of noise power
8.5. Standards for CODEC
8.6. Device minimum capability, including standby mode, reference voltage, and zero circuit
8.7. Basic telecommunications, including types of digitization of audio signals
8.8. CODEC requirements for testing, including idle channel noise, gain tracking, and signal-to-noise ratio
8.9. Introduction to CODEC testing, with explanations of various tests, and a graph of the path of operation in coding and decoding (multitone, full-channel, and half-channel tests, continuity tests).

8.10. Single-tone CODEC testing (analog bench setup method). Detailed description of the most complicated parameter testing in a very simple, analytical and easy to comprehend fashion.
8.11. Multitone CODEC testing; description of the procedure; advantages and disadvantages of the method.
8.12. Full-channel and half-channel testing descriptions.
8.13. Supply current testing in step-by-step procedure.
8.14. CODEC dynamic tests: encoder
8.15. CODEC dynamic tests: decoder

A coder/decoder (CODEC), which converts analog input into digital format (A/D) and digital input into analog format (D/A), is an essential part of a communications network. This device consists of an ADC, a DAC, an internal clock, successive approximation register (SAR), and filtering stages. Although the device is called a coder/decoder, its function is more complex than just the A/D and D/A conversion of received signals.

There are various types of CODECs available and their applications vary as the network demands. The levels of programmability, the clock frequencies and the technologies used in their fabrication represent the main differences. A CODEC converts frequencies between 300 and 4000 Hz (which includes human speech) into 8-bit digital signals (A/D) at its receiving end, and sends them to a switching unit at its transmit end in serial sequence. The receiver portion of the CODEC controls the conversion of 8-bit signals as they arrive serially.

The CODEC also coordinates the timing between itself and the networks with which it interfaces. Data transmission is synchronized and multiplexed. When the encoding section receives the analog signal, the signal is passed through filtering stages to suppress low-frequency noise and the anti-aliasing process is applied to it. Subsequently, the filtered signals are sampled at a defined sample rate per second by an A/D converter and then digitized. The result is a sequence of 8-bit digital words that are equivalent to the original analog signal. Digital signals received by the CODEC are processed in reverse order. They go first to the pulse code modulation (PCM) unit and are then filtered before being converted (D/A) to an analog, audio frequency signal. The sampling rate is the same as in the A/D process.

A process called *companding* (compression and expansion) is essential to the CODEC's performance. Digitization of the analog pulse is not uniform, because input pulses themselves are not uniform. Companding is used to upgrade the quality of the pulse, which increases the signal-to-noise ratio and reduces peak power to prevent overloading. Both the encoding and the decoding sections of the CODEC perform the companding process. The internal device controller controls the process and keeps track of the status of related switches. The function of a CODEC is the reverse of a modem.

8.1 Frame Concept

In data communication, the information about the supply address, call initiation, and the setup and termination of the connection are outlined in

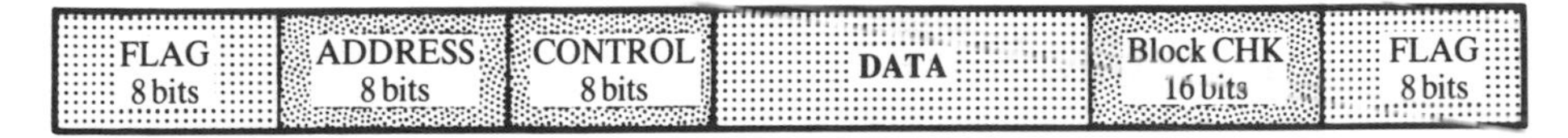

Figure 8.1 General structure of a frame.

"protocols". The three main protocols are

- The American National Standard Institute (ANSI)
- The International Standard Organization (ISO)
- The Synchronous Data Link Control (SDLC) by IBM

The main function of protocols is interfacing between the communication networks and the host computer. The frame, which is the basis for the data transmission unit, has the common format depicted in Figure 8.1. It is composed of flag, address field, control field, block check, and the final or ending flag.

All protocols use the same standard. The structure of all frames is the same except for the "information" (data) section, whose length varies.

- The 8-bit flag sequence at the beginning and the end of every frame is standard (01111110) to maintain synchronization.
- The 8 bits in the address section enable the system to have access to another 256 stations or groups of stations.
- The sequence of binary digits in the control section can be extended as desired. Increasing this sequence provides different frame sequence numbering.

The AT&T μ-255 law (also called the mu-law) outlines the regulation of data transmission. The curve of the process is linear because $\mu = 0$ and increasing. In the μ-255 law, the encoding process is piecemeal and linear.

Once encoded samples enter PCM, they can be used in an application as needed. To deal with noise that critically affects the performance and testing of CODECs, the device usually has two separate grounds for the internal filtering of analog and digital networks.

Some CODECs are combined with a filter in a separate chip that requires additional wiring connections to the main device. This CODEC/filter combo is preferred in certain applications. Whichever CODEC is used, it must be able to drive a 600 Ω resistive load.

Each frame contains 24 channels formated into one word each at a rate of 8000 samples per second. CCITT regulations specify the structure of a frame. Each frame:

- is one word long;
- has the same sampling rate throughout;
- has 256 bits (0 to 127 for positive and 0 to -127 for negative signals);
- has 2048 million bits transmitted per second;
- has one bit containing signal information for dialing, alarm, ringing, off-hook, etc., on the telecom line

The 8-bit code representing the signal is divided into three segments with a flag bit (either 0 for negative or 1 for positive).

There are a total of

$$(24\,\text{words} \times 8\,\text{bits}) + 1 = 193\,\text{bits}$$

This formula is based on a 24-word frame. An extra bit (+1) is used as a unique marker, to identify each frame.

In A-law: The width of every chord (see Figure 8.2: the space between each pair of dots is called a chord) is double the preceding one, after the first two chords, which are the same width.

In μ-law: The width of each chord is double the width of the previous one as they move away from the origin in any direction along the curve. Every step also doubles in width every 16 steps.

8.2. Companding Rules in Communication

To obtain a uniform signal-to-noise ratio over a wide dynamic range, there are two kinds of logarithmic rules for companding:

1. *μ-Law* is the standard set accepted by North America and Japan:

$$Y = \frac{\ln(1 + \mu|x|)}{\ln(1 + \mu)}$$

where

Y = coding range

x = amplitude, $-1 \leqslant x \leqslant 1$

The value of μ is determined by the law and is constant.

2. *A-Law* is set and accepted by Europe. The main difference between A-law and μ-law is that in A-law the normalized amplitude of x is replaced by a signal whose amplitude is smaller than $1/A$. The normalized expression for the A-law is

$$Y = \frac{1 + \ln(A|x|)}{1 + \ln A} \qquad \text{where } 1/A \leqslant |x| \leqslant 1$$

$$Y = \frac{Ax}{1 + \ln A} \qquad \text{where } 0 \leqslant |x| \leqslant 1/A$$

The value of A is determined by the law and is constant.

A minor difference between the two is in the variation of signal-to-noise ratio, that for the A-law being slightly higher. Schematics of the μ-255 and A-87.6 laws are given in Figures 8.2 through 8.5. There are 15 segments on each curve, 7 on the positive and 7 on the negative side, and another segment in the zero area.

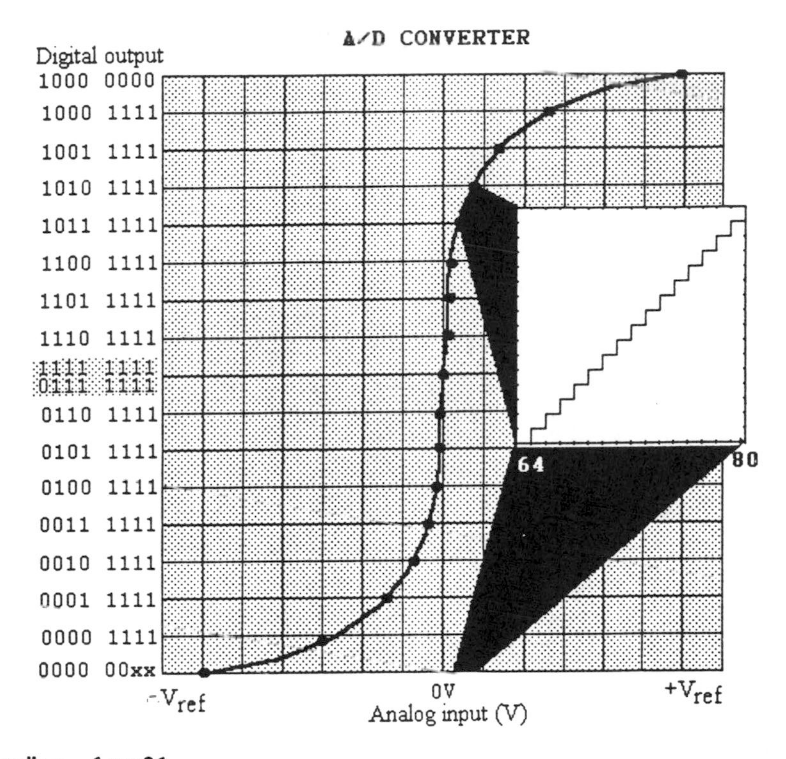

Signaling 1.xx 01
Signaling 0.xx 10
No Signaling .xx 10

Figure 8.2 A/D converter (encoder) 255 (micrometers)-law quantizing curve.

Each segment consists of 16 steps of equal size. The size of a step is always twice the size of the previous step (refer to Figure 8.6).

The expanding operation occurs in very small phases at 0 V, as shown in Figure 8.2. Phases enlarge and compression results when the analog input voltage increases. The D/A μ-law companding process transfer function is shown in Figure 8.4.

	μ-law	*A-law*
Dynamic range	78 dB	66 dB
Constant signal-to-noise	40 dB	35 dB

Whichever law is used, the aim of companding rules is to guarantee a quantization ratio proportional to the size of the original signal.

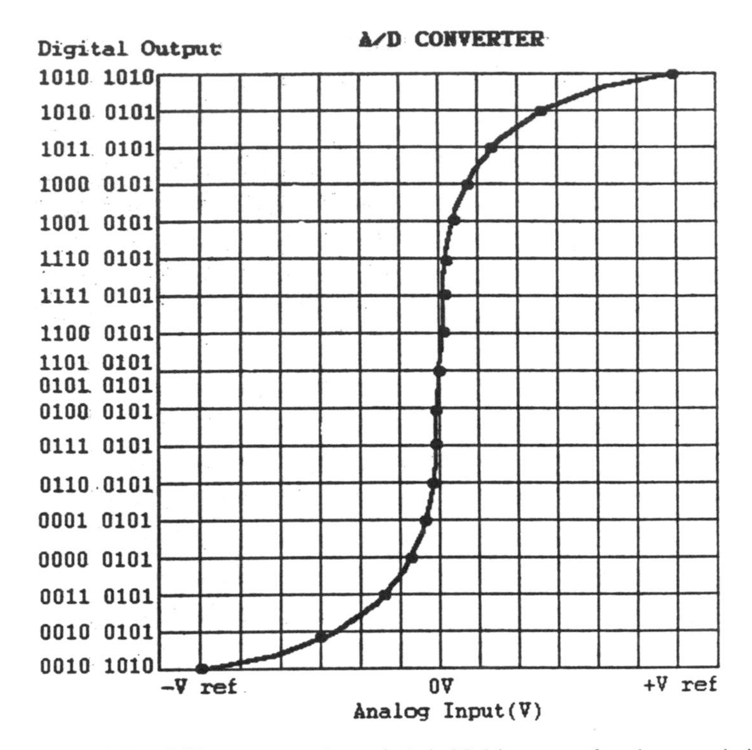

Figure 8.3 A/D converter (encoder) A-87.6-law transfer characteristics.

8.3 C-Message and P-Message

The most important noise affecting CODEC design is idle channel noise (ICN), which exists even when there is no signal traffic. ICN occurs as a result of the quantization of a 0 V input analog signal as the digital output bounces around the assigned binary value for zero along with any nonlinearity existing along the transmission line.

Although some manufacturers' device data sheets give CODEC noise limit specifications, the information is not in commonly accepted units such as *C-message* and *P-message*, which are psophometrically weighted, universally known weights for noise identification. Both are weighting curves that represent the human ear's response to noise. The curve shows the level of annoyance. Psophometric or weighted noise is a measuring set used to measure the level of noise in ordinary commercial telephone networks. The CCITT criteria for the measurement of crosstalk or any other interference between two adjacent lines in communication systems are as follows.

A uniform spectrum random noise produced by an external generator is passed through a weighted network. This random noise, which is "white noise," has an effect on the adjacent channel as crosstalk. The crosstalk is then measured with

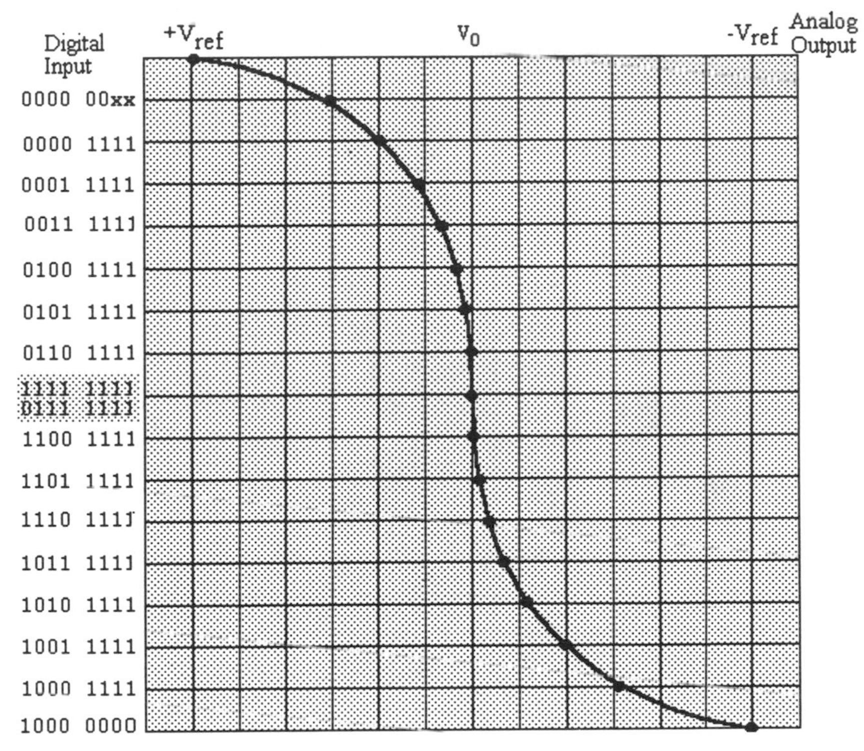

Figure 8.4 D/A converter (decoder) 255 (micrometers)-law transfer characteristics.

a psophometer, which is a weighted measurement tool. Telephone system noise is within the passband of speech frequencies (0 to 4 kHz). To identify the level of annoyance, it is most important to know the noise frequency range. C-message is accepted in the United States and P-message is common in Europe; there are only minor differences between them. The C-notch filter is a combination of a notch filter with C-message that has a notch filter around the frequency of 1 kHz. The frequency of the holding tone in this type of filter is 1004 Hz to avoid interference with pulse code or other such modulations. In the center of the voice frequency band, the difference between the two messages is less than a decibel. Both messages are shown in Figure 8.7.

8.4 Unit of Noise Power

Since the power of noise in a CODEC is very small in terms of ordinary decibels (dB), a more appropriate unit is chosen for this purpose. The following

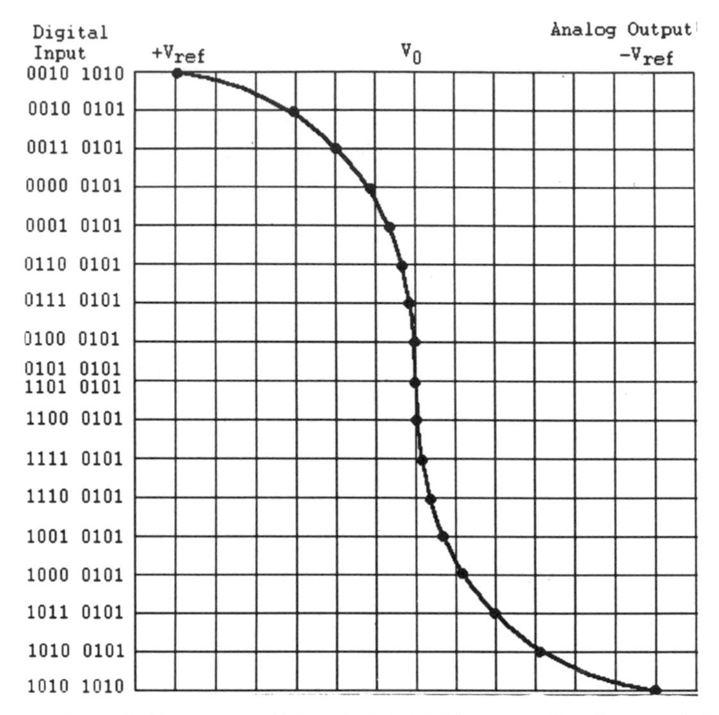

Figure 8.5 D/A converter (decoder) A-87.6-law transfer characteristics.

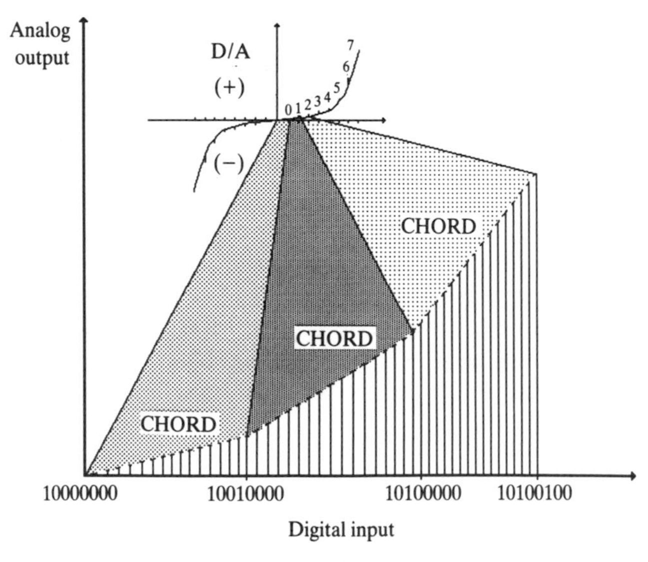

Figure 8.6 Decode characteristics in D/A conversion.

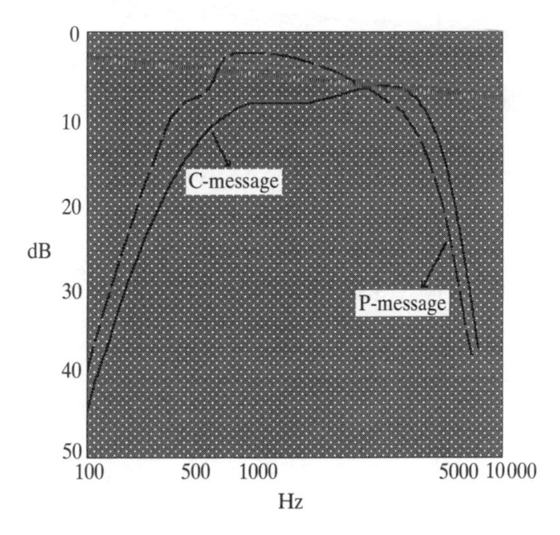

Figure 8.7 C-message and P-message on logarithmic scale.

are commonly used in CODEC noise power measurement.

- dBm unit for noise referenced to one milliwatt (dBm = 10 log[power/1 mW])
- dBm0 unit for noise power measured at zero transmission level point (0TLP)
- dBm0p unit for noise power in dBm0, measured by noise measurement having C- or P-message weighting
- dBrn unit for noise power above reference point, a zero dBrn is equalent to −90 dBm
- dBrnc unit for noise power in dBrn, measured by noise measurement having C-message weighting
- dBrnc0 unit for noise power in dBrnc measured at zero transmission level point
- dBW, referenced to one watt
- dBrn or dBp unit for noise power referenced to one picowatt (rn = dBm0 − dBr)

Example If a device data sheet states that the nominal input level of 3 dBm produces 0 dBm0, and 12 dBrn of noise power has been measured, then the value in dBrnc0 is

$$12\,\text{dBrnc} - 3\,\text{dB} = 9\,\text{dBrnc0}$$

where

n = reference noise
c = measurement through C-message
0 = 0TLP (transmission level point)

8.5 Standards for CODEC

There are minimum standards for all CODEC devices that include noise immunity and distortion. Two important standards are

1. AT&T in D# and D$ design objectives
2. CCITT in G.711, 712 and G733

Any CODEC design must incorporate one or both of the above standards. The system end-of-life standard is required by the US Bell systems.

8.6 Device Minimum Capability

All CODECs must have at least the following capabilities.

8.6.1 Standby Mode

This is the ability to disable the device and reduce its power consumption by a predefined amount, sometimes called a power-down. The device is disabled by loading a control word with a "1" bit in the first two locations. This mode powers down all the functions of CODEC except the data intput for device programming, clock, signaling input and output to and from the frame, and the input to the power-down (the one TTL load can still be driven by the device).

8.6.2 Reference Voltage

The range of a device's operation is discussed in Chapter 6.

8.6.3 Auto Zero Circuit

A circuit created by connecting an *RC* time-constant circuit to the sign bit of the A/D converter. If the *RC* timing is long enough, the DC offset of the incoming signals will be canceled.

8.7 Basic Modulations in Telecommunication

Because digital signals can be manipulated much more quickly and easily than analog signals, the digitization of audio signals has grown rapidly. Many manipulation methods have been assigned, including:

PWM Pulse width modulation
PAM Pulse amplitude modulation
DM Delta modulation
PCM Pulse code modulation

Pulse code modulation processes in North America and Japan include sampling, quantization, and signal coding. The first of two filtering stages that an incoming analog signal (100 to 8000 Hz) passes through results in an output signal with a cutoff frequency of 3400 Hz. The second stage is for anti-aliasing. After filtering, the signal is ready for sampling and digitization. It is sampled using the Nyquist rate then digitized into 8-bit words by an A/D converter. Sign and magnitude bits in these words are transmitted serially to the application network.

8.8 CODEC Requirements

There are three important parameters to be considered and tested in verifying the integrity and performance of a design:

1. Idle channel noise
2. Gain tracking (linearity)
3. Signal-to-noise ratio

8.8.1 Idle Channel Noise

The test for idle channel noise is done at the CODEC output (on the A/D converted signal) with a 600 Ω load, and measured in decibels. While the analog input is shorted, the digital output response is collected and an RMS operation is performed on the data. Inputs to the decoder are digital quiet codes, all "1's".

8.8.2 Gain Tracking

Another name for gain tracking is linearity. It concerns the ability of the output to linearly track the input power level and show how nonlinear A/D and D/A affect each other. For this test, known frequency and input voltage are varied and the ratio of amplitude distortion over input voltage variation is measured.

8.8.3 Signal-to-Noise Ratio

Signal-to-noise is also called C-notch and is in the same family of tests as harmonic distortion measurement. This measurement is the ratio of signal power to the noise power as a result of the quantization process. A known fundamental frequency is applied to the device and the resulting noise distortion is measured by a C-notch filter internal to the mixed-signal ATE.

8.9 Introduction to CODEC Testing

From the previous pages we have seen that a CODEC device contains many complex functions and several sections. These include amplification, sampling, quantization, encoding, and signal reconstruction. Any test suggested by the manufacturer must cover all customer requirements. Because of the complexity of a CODEC device, most of the required tests are lengthy and time-consuming.

Harmonic distortion, signal-to-noise ratio, and noise measurement are the most complex, requiring special techniques and sophisticated circuitry.

There are two sets of tests (full-channel and half-channel) for a CODEC device depending on whether its function is A/D or D/A. The encoding or decoding sections of the device require numerous specific tests such as the following. To test the D/A function, a sequence of synchronized digital bits can be applied to the input port by a mixed-signal ATE and the analog response measured at the analog output port. The system can measure the digital response at the digital port after application of audio signals to the analog input port to test the encoding section. These tests require mixed-signal capability of the test system.

The functioning of the CODEC device is much more diverse than A/D and D/A conversion, however, and more complex tests are needed to verify all its functions. It is best to perform two types of measurement, one that demonstrates the specificity of the design and one to evaluate all of the device's capabilities. These two types of tests are lengthy and extensive. It is not unusual to have hundreds of tests for each A/D and D/A section of the device. Some test results can never be measured accurately due to the lack of exact correlation between the mathematical formulas and the functions of the corresponding electronic hardware.

The signal-to-noise ratio test measures the ratio of signal power to the noise power that results from the quantization process. This test is in the same family as harmonic distortion measurement and is also called C-notch. The noise distortion, resulting from applying a known fundamental frequency to a device is measured by a C-notch filter, which is controlled through the user software codes.

Fortunately, digital signal processing based on Fourier series and transforms, as described in Chapter 7, is the key to assisting test engineers with these complex measurements. Figure 8.8 shows schematically the path of operation in coding and decoding of a 16-bit CODEC. In a complete communication circuit (e.g., telephone line), two CODEC devices are required for a two-way conversation.

The functional sequences (patterns) as well as the timing sequence generated by the ATE must be accurate, especially for production testing; accuracy is the key to guaranteeing repeatability. This is especially important for some measurements, such as distortion and signal-to-noise ratio.

One of the best sources from which to acquire more information on CODEC testing is *DSP-Based Testing of Analog and Mixed Signal Circuits* by Mathew Mahoney (chapter 12), published by IEEE.

8.10 Single-tone CODEC Testing (Analog Method)

In this section, a description of the most involved parameter measurement in CODEC testing is presented. The connections and the components involved in the test configuration give the reader insight into how the DSP module internal to any mixed-signal ATE is arranged (refer to Figure 8.9). Parameters such as signal-to-noise ratio (SNR), total harmonic distortion (THD) and the gain error are discussed at length.

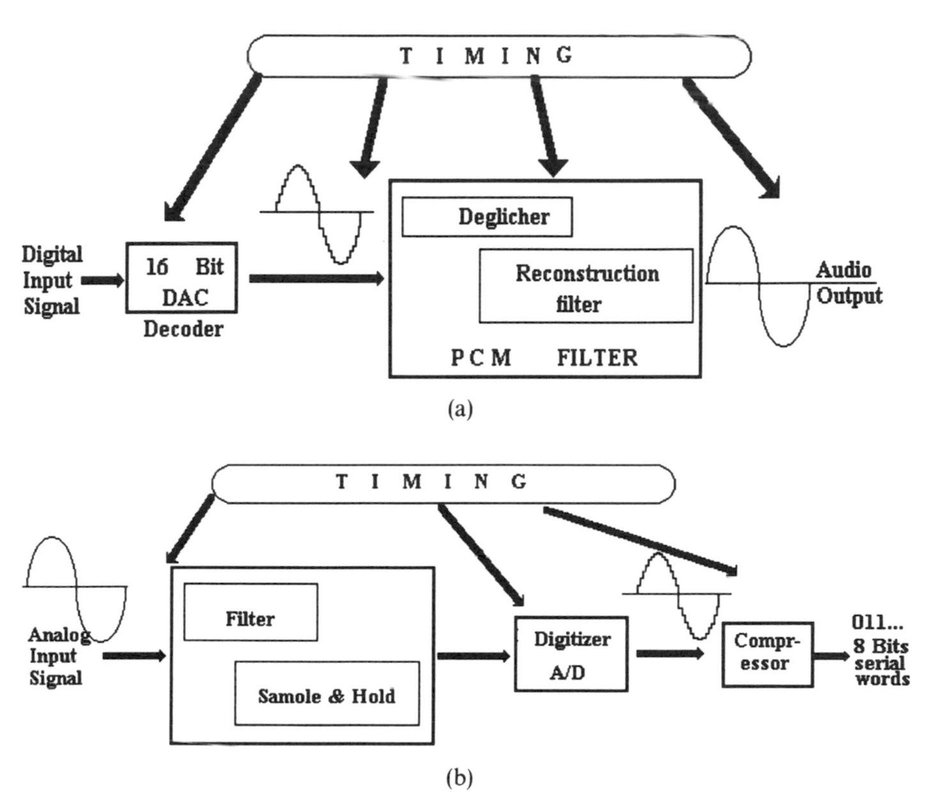

Figure 8.8 (a) Digitized signal reconstruction process. (b) Digitization process.

In contrast to multitone testing, only one optional signal within the band limit of the device operation is chosen as an input and the response of the device is monitored and measured.

Bench testing of a complicated device such as CODEC requires special care in designing test hardware. The ground system must be appropriately considered before soldering any component to the test board. The sampling rate must be chosen with respect to the frequency of the input signal. Each A/D or D/A section of the CODEC has its dedicated timing signals. Without the proper timing signal, proper measurement will be impossible.

Single-tone or multitone CODEC testing is a quick procedure to prove the total functionality of the device in two passes. In the first pass, a sequence of digital codes is applied to the device input buses and the reconstructed output signal is analyzed. In the second pass, a defined sine wave is applied to the analog input and the digitized output signal is reconstructed externally for analysis. In both passes, the internal A/D, D/A, and filter as well as other modules are invoked and examined.

Failures of linearity (including differential and integral), missing codes, and offsets, which are extremely time-consuming, can be detected through the total

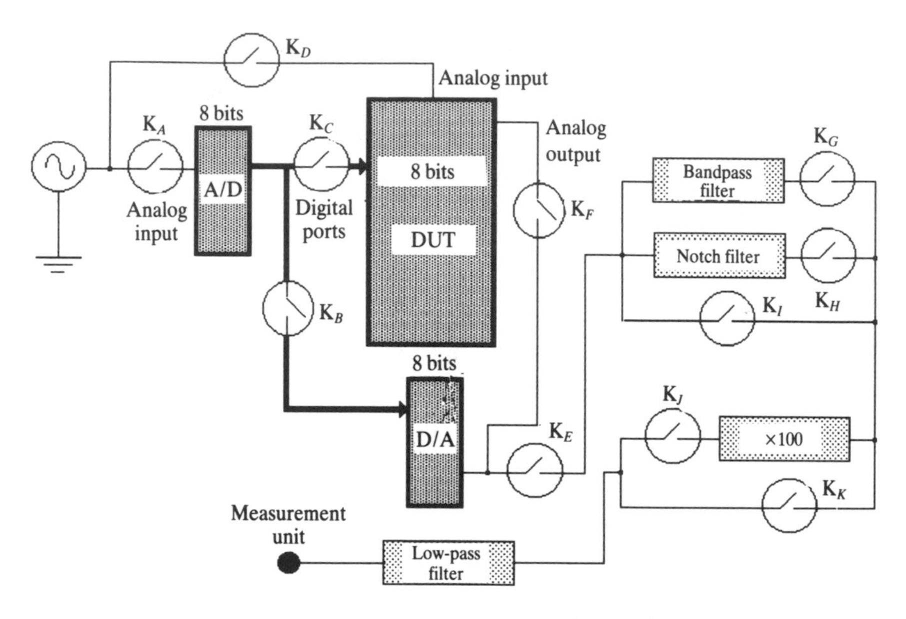

Figure 8.9 Single-tone DAC section testing.

harmonic distortion, signal-to-noise ratio, or gain error tests. Gross functional testing does not reveal information about the faulty section of the device; it is therefore not an appropriate approach for design and engineering analysis.

8.10.1 D/A Section Gross Functionality Test

- Input signal = 2.0 kHz sine wave, 3.0 Vp-p
- Sampling rate = optional (based on the Nyquist rate)

A. To measure the **amplitude of the output signal** as well as the **harmonic (noise)** for reference purpose used in next tests, proceed as follows (referring to Figure 8.9 for notation).

1. Close relays K_A, K_B, K_E, K_I and K_K.
2. Measure the amplitude of the output signal (V_{sig}).
3. Close relays K_A, K_B, K_E K_H, and K_J.
4. Measure the amplitude of the harmonic (V_{noi}).
5. Calculate the percentage of the total reference harmonic distortion as follows:

$$\mathrm{THDR} = \mathrm{ABS}\left[\frac{V_{sig}}{V_{noi}}\right] \times 100$$

B. Gain Error Measurement

1. Close relays K_A, K_C, K_F, K_E, K_I, and K_K.
2. Measure the amplitude of the output signal (V_{err}).
3. Calculate the percentage of gain error as follows:

$$\text{Gain error} = \text{ABS}\left[\frac{V_{err}}{V_{sig}}\right] \times 100$$

4. The measured value must be within the high and low limits.

C. Total Harmonic Distortion (THD) Measurement

1. Close relays K_A, K_C, K_F, K_E, K_H, and K_J.
2. Measure the amplitude of the output signal (V_{ne}).
3. Calculate the percentage of THD as follows:

$$\text{THD} = \text{ABS}\left[\frac{V_{ne}}{V_{err}}\right] \times 100 - \text{THDR}$$

4. The measured value must agree with the maximum and minimum limits.

D. Signal-to-Noise Ratio (SNR) Measurement

1. Close relays K_A, K_C, K_F, K_E, K_G, and K_K.
2. Measure the amplitude of the output voltage (V_{snr}).
3. Calculate SNR as follows:

$$\text{SNR} = \text{ABS}\left[\frac{V_{snr}}{V_{ne}}\right]$$

4. The measured value must be within the maximum and the minimum limits.

8.10.2 A/D Section Gross Functionality Test

- Input signal = 2.0 kHz sine wave 3.0 Vp-p
- Sampling rate = optional (based on the Nyquist rate)

A. The same procedure as for THDR is also applied here (refer to Figure 8.9 for notation).

1. Close relays K_A, K_B, K_E, K_I, and K_K.
2. Measure the amplitude of the output signal (V_{sig}).
3. Close relays K_A, K_B, K_E, K_H, and K_J.
4. Measure the amplitude of the harmonic (V_{noi}).
5. Calculate the percentage of the total reference harmonic distortion as follows:

$$\text{THDR} = \text{ABS}\left[\frac{V_{sig}}{V_{noi}}\right] \times 100$$

B. Gain Error Measurement

1. Close relays K_D, K_C, K_B, K_E, K_I, and K_K.
2. Measure the amplitude of the output signal (V_{err}).
3. Calculate the percentage of gain error as follows:

$$\text{Gain error} = \text{ABS}\left[\frac{V_{err}}{V_{sig}}\right] \times 100$$

4. The measured value must be within the maximum and minimum limits.

C. Total Harmonic Distortion Measurement

1. Close relays K_D, K_B, K_C, K_E, K_H, and K_J.
2. Measure the amplitude of the output signal (V_{ne}).
3. Calculate the percentage of TDH as follows:

$$\text{THD} = \text{ABS}\left[\frac{V_{ne}}{V_{err}}\right] \times 100 - \text{THDR}$$

4. The measured value must agree with the specified limits.

D. Signal-to-Noise Ratio (SNR) Measurement

1. Close relays K_D, K_B, K_C, K_E, K_G, and K_K.
2. Measure the amplitude of the output signal (V_{snr}).
3. Calculate SNR as follows:

$$\text{SNR} = \text{ABS}\left[\frac{V_{snr}}{V_{ne}}\right]$$

4. The measured value must agree with the specified limits.

8.11 Multitone CODEC Testing

Determining the frequency response of DUTs, usually telecom and audiofrequency converters, is the primary application for multitone frequency tests. Using this method, testing device responses are much faster and more accurate than in the single-tone test. The fast Fourier transform is used once, and the response of the DUT is stored.

In the multitone method, a single input signal, containing all the desired frequencies, is applied to a communications device to determine the device's response to several frequencies. The amplitude and the frequencies of the single stimulus are determined by the engineer through software control.

A frequency domain spectrum containing the signals the test engineer chooses, including random phases, is produced by a generator designed for this purpose. The frequencies of the components of this spectrum are not harmonically related. Since the density function and peak/RMS ratio are dependent on the phase characteristics of the spectrum, the engineer must attempt to reach the desired waveform.

The multitone spectrum contains analog signals with frequencies within those of the passband for telephone communication voice systems (300 to 4300 Hz). These frequencies delineate the minimal performance parameters mentioned previously that are required by American Telephone and Telegraph (AT&T) and the International Telegraph and Telephone Consultative Committee (CCITT) regulations.

The manufacturer's data sheet that is furnished with a telecom device usually states whether the device complies with these requirements.

8.12 Full-channel and Half-channel Tests

The decoding and encoding channels of the device are tested separately, and this is called *half-channel testing.* This method fits production better since the failure in each channel can be verified. It is possible to test the functionality of the device by testing encoding, decoding, and filter in one pass. The only problem is that some of the devices that pass this test may fail in the half-channel testing method.

8.13 I_{CC}, I_{DD} Test (Supply Current)

Purpose

To measure the amount of current entering the device power pin when different operating modes are selected.

Procedure

1. Force V_{CC}/V_{DD} to the power pins and 0 V to GND pin(s).
2. Run clock at the recommended frequency.
3. Choose the mode of operation by setting the related pin.
4. Measure current at the power pin(s).
5. Change the mode of operation (in Step 3).
6. Measure the current through the power pin(s).
7. Test the measured current against the predefined limit.

8.14 CODEC Dynamic Tests—Encoder Section

The encoder section tests consist of the following.

8.14.1 AC Test Gain Tracking (Linearity) Test

Definition

The measurement of deviation of gain on each channel when the input signal level is changing.

Procedure

1. Apply the defined voltage to power supply pin(s) and 0 V to the GND pin(s)
2. Set the function generator to generate the required reference signal with defined frequency and dBm0
3. Apply the generated signal to the encoding input pin of the device.
4. Change the applied signal to different levels in different steps for the specified total number of tests as described in the following instructions.

- **Instruction 1.** In this test samples are collected and sampling is done. The clock frequency in Step 2 determines the number of cycles for sampling, so linearity can be verified in this stage.
- **Instruction 2.** In the above setup, the A-law is applied and the data is converted from PCM code to the equivalent 2's complement. The result is sent to an array processor. The quantization process creates distortion in the reconstructed signal no matter which law is employed. The total noise amplitude with respect to the real signal should be measured. The result will be the ratio of the amplitude of noise to the amplitude of the desired signal. The Fourier transform is applied, and the result will be the *power spectral density*. Based on the unsymmetric created power spectrum, different harmonics can be measured.
- **Instruction 3.** For 0 dBm0 the magnitude of the fundamental and higher harmonics are retrieved, and for other measurements than 0 dBm0 the fundamental values are then considered. The idle channel is created when there is no signal to the input channels of the device. Any noise that is seen at the analog channels is that created by the encoder section of the device during the sampling process and is converted by the decoder to analog noise signals. The best method of minimizing or reducing noise at idle mode is not to ground the input but rather to keep it awake by a minor, nonsignificant signal. Because of the rolloff of the receiver, all harmonics beyond the third are usually zero.
- **Instruction 4.** The array processor that contains the notch and psophometric filters can compute the combination of noise and distortion. *Total noise* can be measured at this stage.
- **Instruction 5.** The signal-to-noise ratio can be measured knowing the result in Instruction 4.

8.14.2 Idle Channel Noise Test

Definition
The noise measured at analog channels when the device is idle.

Purpose
To verify the amount of noise that is created during the sampling process by the device encoder.

Procedure

1. The same setup as for the gain tracking test is also required for this test.
2. Connect a specified load to the device analog input pin to GND.
3. All other steps are the same as for the previous tests. Only the psophometric filter is applied. The remainder will be the total noise when the device is idle.

8.15 CODEC Dynamic Tests—Decoder Section

The decoder section tests consist of the following.

8.15.1 AC Testing

Tests for this section of CODEC are in principle the same as for the encoder section, with the following differences.

1. Digital PCM code equivalent to a sine wave with defined frequency is applied to the decoding input of the device. This sinusoidal wave is already calculated by the CPU based on the data present in the buffer.

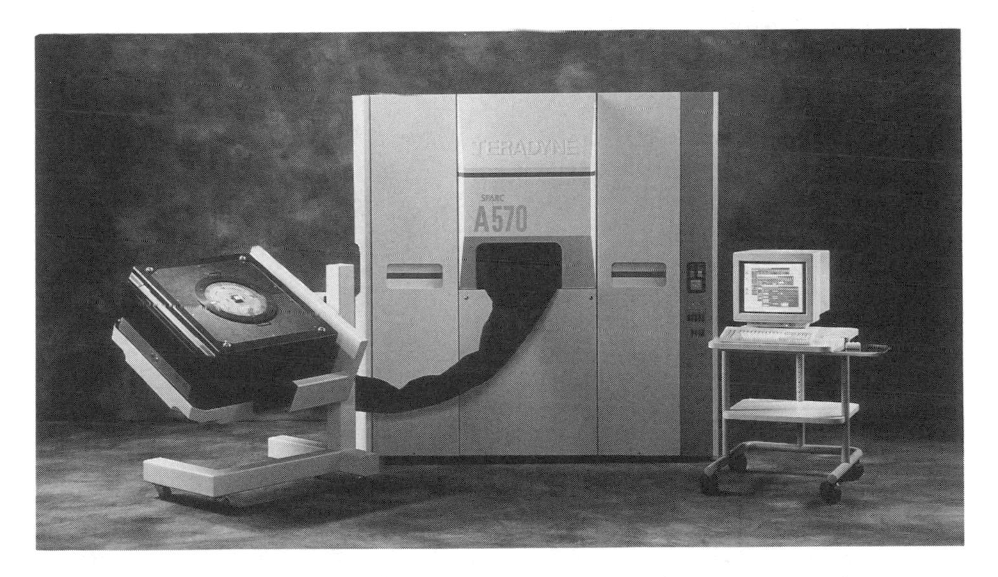

Figure 8.10 The most recent members of the A500 linear and mixed-signal test family belong to the A570/A580 advanced mixed-signal test series, which was introduced in 1992. The A570/A580 series provides up to 192 digital pins with data rates to 200 MHz, edge placement accuracy to ±350 picoseconds, and analog frequencies to 4 GHz for testing a wide range of standard and custom mixed-signal ICs aimed at applications such as notebook computers, digital wireless communications, audio, and graphics and imaging. (Courtesy of Teredyne, Inc., Boston, Mass.)

2. For linearity, signal-to-noise ratio, and harmonic distortion tests, the decoder section must be activated by digital codes equivalent to the defined signal level for the defined number of tests (the same as in the previous section) with the same size of step in each test.
3. The difference between the idle channel measurement in the decoder and that of the encoder is that, in this test, the quiet codes (all "1") are applied to the decoder input.

References

1. N. B. Jones (ed.), *Digital Signal Processing*, IEEE Control Engineering Series 22, Peter Peregrinus, 1982.
2. Mathew Mahoney, *DSP-Based Testing of Analog and Mixed-Signal Circuits*, The Computer Society and the IEEE Press, 1987.
3. Douglas K. Shirachi, *CODEC Testing Using Synchronized Analog and Digital Signals*, Zehntel Automation Systems, EH0237-8/85/0000/0371$0.100, IEEE, 1984.
4. Thomas Young, *Linear Systems and Digital Signal Processing*, Prentice-Hall, 1985.
5. Samuel D. Stearns, *Digital Signal Processing*, Hayden, 1985.
6. Richard A. Roberts and Clifford T. Mullis, *Digital Signal Processing*, Addison-Wesley, 1987.
7. I. T. Young, J. Biemond, R. P. W. Duin, and J. J. Gerbrand, *Signal Processing III, Theories and Applications*, 3rd European Signal Processing Conference, North-Holland, 1984.
8. William D. Stanley, *Digital Signal Processing*, Reston, 1984.
9. William D. Stanley, Gary R. Dougherty, and Ray Dougherty, *Digital Signal Processing*, 2nd ed., Reston, 1984.
10. *Telecommunication Circuits Data Book*, Texas Instruments Inc., 1986.
11. *Linear Circuits, Data Acquisition and conversion*, Vol. 2, Texas Instruments, Inc., 1986.
12. Bernard Sklar, *Digital Communication Fundamentals*, Prentice-Hall, 1988.
13. Robert M. Gagliardi, *Introduction to Communication Engineering*, 2nd ed., Wiley, 1978.
14. James Martin, *Telecommunications and Computers*, Prentice-Hall.
15. C. Britton Rorabough, *Communication Formulas & Algorithms*, McGraw-Hill, 1990.
16. George E. Friend, John L. Fike, H. Charles Baker, John C. Bellamy, and Gerald Luecke. *Understanding Data Communications*, Howard W. Sams, 1988.
17. J. U. Gordon Pearce, *Telecommunications Switching*, Plenum Press, 1981.
18. Bertil C. Lindberg, *Troubleshooting Communications Facilities*, Wiley, 1990.
19. Dr. Kamilo Feher/Engineers of Hewlett-Packard, *Telecommunications Measurements, Analysis, and Instrumentation*, Prentice-Hall, 1987.
20. John A. Kuecken, *Talking Computers and Telecommunications*, Van Nostrand Reinhold, 1983.

Index